A Guide to... Neophema & Psephotus Grass Parrots

By Toby Martin

Published and Edited by Australian Birdkeeper ©

First Published 1989 by
Australian Birdkeeper P/L
PO Box 6288
South Tweed Heads.
NSW. 2486. Australia

Reprinted April 1990

ISBN 0 9587455 1 X

Front Cover:
Top left: Pied Red-rumped Parrot.
Centre left: Scarlet-chested Parrot.
Centre right: Cock Mulga Parrot.
Btm far left: Cream Bourke Parrot.
Btm left: Cock Bourke Parrot.

Design, Type and Art: Jackson Studios (Gold Coast)
Colour Separations: Haighs Foto Art (Brisbane)
Printing: Prestige Litho (Brisbane)

Contents

Introduction to Neophema Grass Parrots .... Page 8
Management .... Page 8-15
Housing; Feeding, Breeding; Hybridising; Ringing; Keep ing Records; Surgical Sexing.
Bourke Parrot. .... Page 16
Turquoisine Parrot. .... Page 20
Scarlet-chested Parrot. .... Page 24
Elegant Parrot. .... Page 28
Blue-winged Parrot. .... Page 29
Rock Parrot. .... Page 33
Orange-bellied Parrot .... Page 35
Introduction to Psephotus Grass Parrot .... Page 40
Management. .... Page 40
Housing; Feeding; Breeding;
Red-rumped Parrot. .... Page 43
Mulga Parrot. .... Page 47
Blue-bonnet . ... Page 49
Hooded and Golden-shouldered Parrots .... Page 53
Paradise Parrot .... Page 58
Diseases of
Neophema and Psephotus Grass Parrots. ... Page 61-79
Introduction; Health Check; Quarantine; Stress & Dis ease; Recognising Disease; Emergency Procedures; First Aid; How to Get the Best out of your Avian Veterinarian; Control & Prevention of Disease; Worms & Worming; Bird Weights & Dose Rates for Worming; Administration & Use of Drugs; Use of Crop Needle; Common Diseases; Diseases from outside your aviary; Prevention of Disease; Disease control; Baby Bird Diseases; Handrearing; Surgical Sexing.
Hospital Cage and Brooder .... Page 76
Use and Design
Handrearing. .... Page 77
Technique & Recipes
Genetic Tables .... Page 78
Publishers Note .... Page 80

Acknowledgements

We would like to take this opportunity to thank Ian Brown, Bill Boyd and Walter Boles, Curator of the Ornithological Section of the Sydney Museum, for their time and effort in assisting in the compilation of this book.

Special thanks to Dr Rob Marshall for his very informative section on disease prevention and cures and to Peter Brown for the chapter on the endangered Orange-bellied Parrot.

Sincere gratitude to Kylie Walshe and Boyd Jackson of Jackson Studios, Gold Coast for their tireless patience and fortitude in the design and production of this book.

Photographic Credits

All Photographs by **Wilson Moore** except for:
Ian Brown: Page 11: centre, Page 27: Top, Centre right, Page 30: Top right, Page 43: Btm left, Btm right, Page 46: Top right, Page 47, Page 48, Page 50, Page 51: Btm left, Btm right, Page 58: All shots, Page 59: Top left, Top right, Btm left, Btm right.
David Heap: Page 18: Centre left, Btm left, Page 19: Centre right, Centre left.
Allan Birrell: Page 22: Top right, Top left, Page 46: Top left.
Gordon Dosser: Page 19: Top right, Top left.
Dave Watts: Page 35, Page 38.
Tod Osborne: Page 27: Centre left.
Alan Chalmers: Page 46: Btm right, Page 54: Centre right.
Bill Boyd: Page 51: Top right.
Gerry Rutte: Page 23: Centre right.
Kevin Wilson: Page 42: Centre.
Tony Silva: Page 46: Top centre.

About This Book

This book is full of information to allow you to get as much pleasure from these birds as possible. Nobody likes losing their birds, everybody enjoys breeding them. Unless the birds are happy both in their environment and in themselves, they will not breed successfully. The details you can glean from this book should be put into practice. Some vital points will be repeated throughout the text.

This book has been written for aviculturists. It concerns birds that have been bred in aviaries and therefore details of their distribution or activities in the wild are minimal.

Every method of breeding these birds are by no means covered by this book. However, the methods discussed have all been very successful. A successful method must be more beneficial to the reader than numerous possible alternatives.

Aviary-bred parrots should be viewed differently to wild parrots, in the same way as the aviary Budgerigar and domestic chicken are to their wild counterparts. Only aviary-bred birds should be considered for aviaries. There is no need for aviculturists to be seen as a threat to existing wild populations. Domestic dogs are not crossed with dingoes or wolves for any reason, neither are domestic rabbits bred with wild ones. The same should apply to aviary-bred birds. Having said that, such birds are now our responsibility and it is up to us to learn how to provide the best for these lives dependent on our care.

This book has tried to avoid the ambiguous, over-generalised statements often seen in other publications. Read, enjoy and learn. Your birds are depending on you.

In Australia all birds discussed in this book are protected under law enforced by each State's or Territory's National Parks & Wildlife Service. Their advice should be sought if you are unsure of the procedure to follow before acquiring these birds.

The Author

Toby Martin was born into a family of bird breeders in Sydney, Australia in 1927. His years were spent working as a shipwright on the Sydney waterfront and travelling the world as what could best be described as an adventurer dealing in shells and all sorts of rare and unlikely objects. He is also quite an artist, specialising in painting the older sailing ships.

He married in 1950 and after twenty-two years his wife found his passion for birds too much to bear. Some years after his divorce, he met his present wife, Jacki, who also bred birds. It was with her that he really specialised in Neophemas and a few of the Psephotus Grass Parrots. He now keeps many of the mutations available today and of course the colours that will combine to produce the mutations of the future.

Toby lectures at many avicultural meetings all over the country. He also founded, and is the current President of The Neophema and Grass Parrot Society of Australia.

A Personal Note from the Author

I would just like to say here that I suffer from dyslexia. I have great difficulty reading and writing. This book is an example of what can be achieved through determination and the assistance of friends. There are many experienced aviculturists who have a wealth of knowledge to share but for one reason or another can't put it down on paper. I would strongly urge such people to contact "Australian Birdkeeper" who will do all they can to assist, as Wilson Moore has done for me.

Toby Martin

The Neophema Grass Parrots

Introduction

These pretty little parrots have everything going for them. They are attractive, relatively quiet and peaceful, and can be mixed with finches and other non-aggressive parrots like the Princess Parrot. Being small they do not impose on us the building of large aviaries as other larger parrot and cockatoo species do. With a few exceptions they are usually willing to breed.

They have a very reasonable lifespan. A cock Turk is still alive today at 17 years of age and a hen Bourke will be breeding this year at the age of 10.

Some aviculturists will be pleased to know there are now quite a number of Neophema mutations. Some are available now, others are still relatively rare at this moment in time. The mutations will be discussed under the relevant bird.

They are definite family favourites - and gentlemen, these are often the birds that are the key to success when trying to persuade your wife to move the washing line a bit to the left in order to accommodate a few more aviaries.

The seven members of this family are:

Bourke's Parrot *Neophema Bourkii*
Turquoisine Parrot *Neophema Pulchella*
Scarlet-chested Parrot *Neophema Splendida*
Elegant Parrot *Neophema Elegans*
Blue-wing Parrot *Neophema Chrysostoma*
Rock Parrot *Neophema Petrophila*
Orange-bellied Parrot *Neophema Chrysogaster*

Management

Housing

My aviaries for breeding single pairs of Neophemas measure 2.10m long by 1.80m high and 80cm wide. This allows me to get eight such aviaries in an area approximately 6m long. Originally my aviaries were 3.6m long but I have found the Neophemas do well in the shorter flights.

These 2.10m aviaries not only suit the birds but also save me space and some dollars. One of the main points in favour of shorter aviaries for the Neophemas and Psephotus Grass Parrots is that both adult and young birds are capable of picking up speed very quickly, and in long aviaries fatal accidents often occur when contact is made against the often unseen front wire. This is especially the case in young birds who are unfamiliar with their new surroundings. If your aviaries are already quite long, this problem can be lessened greatly by placing soft branches of Tea Tree or pine in front of the end wire.

The shelters are 1.2m long by 80 cm wide and 1.80m high. 45cm high tin divides the shelter from the flight at ground level. This stops vermin entering the shelter. 20cm of tin drops down from the roof to stop rain blowing in. The fibro ceiling is insulated with 7.5cm insuwool below the galvanised roof. The concrete floors have 5cm of Course River Sand over them. (Comments will be made on aviary floors under Worming.)

The fibro walls and ceiling inside the shelters are painted white. I use Acrylic Industrial Flat White paint which is cheap, washable and harmless to the birds. The majority of paints available today do not contain lead. I also have a bit of blue added to the paint because brilliant white is believed to be uncomfortable for the birds. This brightens up the shelter and does not discourage birds from entering. Dark shelters, in comparison with the outside flight, are viewed with suspicion by many birds.

The shelters, afterall, are where I want my birds to breed, feed and generally retreat if disturbed or if the weather is unfavourable - therefore they must be very acceptable to them.

Not shown to the left is a storage area and a food preparation area and sink.

The shelters back on to a 1.80m wide enclosed walkway. This area allows me to set up hospital and quarantine cages and to store general equipment against the opposite wall. They are then conveniently to hand when required. By entering the aviary from the rear, the birds never have to fly at you, so this causes them less or no stress at all. Seed hoppers are attached to the 90cm high back door. They have a wire tray that allows seed and husks to drop into a catching bin below. This makes for a cleaner aviary and also allows the good seed, normally wasted on the ground, to be re-used. If this is done, as a disease prevention, the "recycled" seed should only be used in the aviary from which it is collected. Next to the hopper a wire basket/shelf holds fruit, vegetables, sprouted seed and arrowroot biscuits etc.

There is a fixed perch in the shelter. I use a natural branch of varying thicknesses as this exercises the birds' feet and also lets them choose the most comfortable width on which to roost.

Seed hoppers and 'extras' tray are easily serviced when the door is open. Water bowl is just inside the door.

In the outside flight, however, I use a swinging perch. The ends do not come into contact with the wire sides. This I find helps stop the cock bird from squabbling with the cock next door when he should be paying attention to his hen. In order to squabble they have to hang onto the side wire. They find this awkward and don't really like it. It also encourages them to exercise, especially when keeping balance on landing. Over-fat birds often fail to breed, and this must be watched for.

The aviaries should face North to North-East. The solid backs therefore face the cold southerly winds and the front gets maximum sunshine. Draught and damp are enemies of our birds and they should be excluded from any aviary. A friend of mine once told me that if you have a nail hole or whatever in the side of your shelter, wherever it is a Scarlet would find it and sit there in the draught and get pneumonia - So be warned! I do have one bank of aviaries of the same size completely roofed and enclosed with fibro, apart from the top 90cm at the front. These are excellent for problem or newly acquired birds - especially those from a warmer climate. They are also a good idea if for some reason your aviaries cannot face the ideal direction. Late or early breeders can use these aviaries when the weather tends to be inclement. These aviaries also allow argumentative birds to be housed alongside one another. Unable to see the bird next door, there is less fighting and more breeding. Quite often, because they can hear one another, cock birds of the same species often bring each other into breeding condition.

I also have holding aviaries for young and non breeding birds. These are 6.60m long, 90 cm wide and 2.10m high. The shelter is 2.10m x 90 cm x 2.10m. These aviaries will hold anything up to thirty Neophemas. Soft bushy branches are placed on the end wire giving the young birds plenty of warning to slow down.

A word of warning is relevant here. One morning I went down to my young bird holding aviaries and found odd Bourkes, Turks and Scarlets lying dead on the floor. After inspection, I found they had all died from cerebral haemorrhage. I couldn't work it out. My first thoughts were of cats or owls disturbing them at night. When I informed an old breeder friend of mine, he immediately asked if I was flying Elegants in the same aviary. Yes, I told him. That's your answer he said: go down to your aviary at dusk and you will see the Elegants bombing the others and hitting them on the head with their beaks. This I did and saw it for myself. Since then I have never held Elegants along with other Neophemas.

Feeding

Seed quality is a bit of a worry to me in Australia today. It would appear that aviculture is being used as a dumping ground for below standard seed. To avoid sub-standard seed, I suggest you first find a reliable seed source. Then before buying a full bag or bags of seed, buy one or two kilos. Take it home and soak it for no more than twelve hours. Then spread it out on a cloth and allow it to sprout. (Sprouting jars are available for this job.) Before I buy any more seed 70% - 80% must have sprouted. If only 50% sprouts then half the seed is dead. This is only wasting your money and the birds merely benefit from half of what is given to them.

My basic seed mix consists of 6 parts Canary, 1 part French White Millet, 1 part Hungarian Millet, 1 part Grey-striped Sunflower (I prefer the small sunflower myself) and 1/2 part of Hulled Oats. I buy seed only three or four times a year and store it in 40 gallon drums.

In addition to dry seed, I feed a lot of sprouted seed. I have forty aviaries and in the breeding season they take 5 - 8 litres of sprouted seed per day. I soak Red Pannicum Millet, Grey-striped Sunflower and Mung Beans in separate containers. These are soaked in water for about twelve hours (usually overnight) - never any longer. They are then rinsed and left for day 1 in large sieves to sprout. They are then rinsed again and left for day 2. On the morning of day 3 the sprouts should be almost 1cm long. (Once over 1.2cm the sprout is past its prime and becomes drier.) I then rinse it and store it in a fridge for about six hours prior to feeding.

They then get a final rinse and are soaked for half an hour in an anti-bacterial solution. I then mix all three sprouts into a 5 litre plastic bucket. To this mixture I stir in two heaped teaspoons of vitamin and mineral powder, and for animal protein I mix in four teaspoons of powdered egg yolk. The sprouted seed is then given to the birds between 3.00 and 4.00 o-clock in the afternoon. This I find is important - by feeding sprouted seed in the afternoon, any that is left over will sit overnight and be eaten first thing in the morning. In the heat of the day the sprouts will dry up and be wasted. The young also benefit from a good feed early in the morning. During this afternoon feeding, I also give the birds 1/4 to a full arrowroot biscuit and whatever is available at the time in the way of apples, silverbeet (spinnach), sweetcorn, broccoli and carrots, cut or shredded into manageable pieces. I also check the seed hoppers and attend to the shell grit, cuttlefish and rock salt (I use a horse lick block which is a mixture of Molasses and Rock Salt).

I use white porcelain microwave oven dishes about 30cm square and 5cm deep for drinking water. These are easily cleaned. They are placed on bricks on the ground well away from perches, and fresh water is given daily.

In the morning the birds are given seeding grasses, thistles and dandelions as available.

▲ *For sentimental reasons, Toby still keeps a few Jacarini Finches*

Drinking water is provided in white porcelain microwave oven dishes ▼

In a longish aviary some brush will give advanced warning of the end wire to young birds ▼

Breeding

For me, the Neophema breeding season begins in early to mid July and continues through to the end of March/beginning of April.

All Neophemas perform the same basic courtship display. The cock struts and hops around the hen, bobbing up and down, stretching up to his full height and drooping or raising his wings and fanning his tail while giving his twittering call. His head bobs up and down as he regurgitates food and feeds his hen. Neophemas will breed at 8 - 12 months of age. I prefer hens that are at least 12 months old, and cocks seem to do better if at least 18 months old. Mating will take place on the perch or on the ground. In all Neophemas only the hen broods. I have had no real success breeding Neophemas in trios. One pair per aviary undoubtedly brings the best results. The only exceptions to this are Rock Parrots and Blue-wings where breeding in colonies is much more successful.

The method of breeding I use to great effect is to pair up the birds and allow the hen to have two good clutches - if the clutches are small, I sometimes allow them three. I do stop her if she has reared eight or more babies. I had a hen Bourke rear sixteen young in one season. The following season she only reared two, so don't take too much out of your hens or your following season will be disappointing.

I carry spare cocks and hens in holding aviaries. When I stop the original hen from breeding, I leave the cock bird in the aviary and give him another hen, and she is nearly always desperate to go to nest. I then take two clutches from her, remove the pair and start again with a new pair. The incubation time in Neophemas varies between 18 and 21 days. Many factors can alter this time however - for example, cold weather, when the hen may not start to incubate properly until the third or fourth egg has been laid, and the first three eggs may be infertile etc. So realistically and to be safe, from the time the first egg is laid to the last chick has hatched will be anywhere between 18 and 30 days.

The young stay in the nest between 4 and 6 weeks. When they fledge they should be left with their parents for about 4 weeks. They will obviously have to be removed earlier if being persecuted by the cock bird.

For the most part Neophemas are non-aggressive. I will not deny you won't come across the odd rogue bird. You will just have to watch out for this. Many people think their pairs are incompatible because the cock and hen squabble and bicker between themselves, but I have found these pairs to hatch and rear more young than the really peaceful or docile pairs do.

Quite often you may find that for two seasons in a row you will breed a lot of hens, then in the third season a lot of cocks. This phenomena does happen in the wild as well and I'm afraid I have no explanation for this.

One opinion with which I do agree concerns the theory that hens can control the fertility of their eggs. If in a breeding season the hen sees a lot of rain which she associates with plenty of food (like seeding grasses etc) she may lay six eggs of which four to six will be fertile. In a bad dry season she may not have one fertile egg. I have had numerous hens which have reared eight to ten young for two consecutive seasons. Then we had a dry year and the same hens laid about twenty eggs each and fertility ranged from zero to two. I am sure it is not the cock's fault as they mate normally. The hen just has this "control" over egg fertility.

Most Neophemas generally breed well. Blue-wings are an exception, as one season can be good and the following two dismal. They do fluctuate a lot from year to year. The other exceptions would be some of the newer mutations. To clarify this a bit more I would go as far as to say that out of 25 pairs of normal Neophemas a good season should produce between 120 and 140 young.

From 25 pairs of a well established mutation such as the Red-fronted or Yellow Turks a good season should produce 70 to 80 young.

From 25 pairs of the newer mutations such as Rose, Pink or Cream Bourkes or Cinnamon Elegants if 40 to 50 young were produced then you would have had an excellent season.

From this it can be seen that none of the Neophema mutations are as strong as the normal at present. (More will be discussed on mutations under the specific bird section).

Nest Boxes

At present I am using two types of nest box. The reason for this is that some of my birds prefer the open top boxes and some the dark enclosed ones. Therefore, I give them a choice and then remove the one not being used. I will describe both boxes. My enclosed nest box is 20cm square and 30cm deep. The hole is in the centre about 6cm from the top and is 5.5cm in diameter. A piece of 1.2cm dowelling sits below the entrance hole protruding 7.5cm outside and 6cm inside the box. A length of 5cm x 6mm timber is wrapped in 1.2cm x 1.2cm wire and fixed to the inside front of the box. The hen and the young can easily climb up to and sit on the inside dowel to be fed by the cock through the entrance hole.

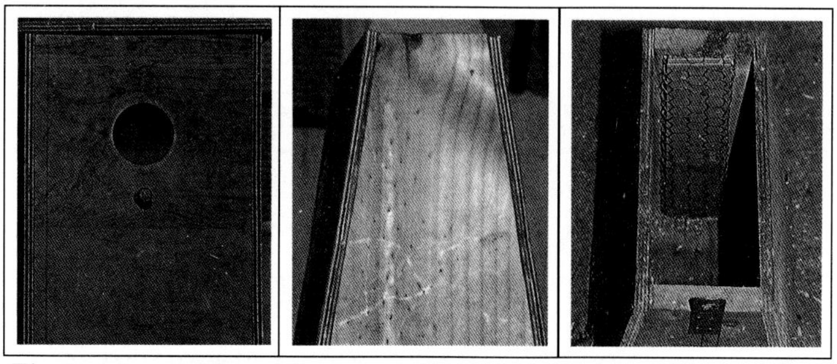

The two types of nest boxes used by the author for his Neophemas. The pyramid has no lid.

In the summer these boxes can sometimes get too hot. This can be alleviated by sliding the lid back or propping it up, leaving a 2.5cm gap to allow hot air to escape. In this condition, attention must be paid to the weather. If you get a cold snap in the late afternoon or at night, the babies can be fatally chilled if the hen is away. So you sometimes have to be opening and shutting lids much of the time. I hang these boxes about 4cm from the roof.

The open model of nest box is again 20cm square and 30cm high but the front and back tapers to a wedge shape at the top. There is no hole; instead the top is open forming a 20cm x 7.5cm slot. The same wire-clad climbing timber ladder is used inside. No dowel is used as the birds sit on the top. This box is hung about 15cm from the roof, allowing easier access by the birds.

These boxes hold the warmth at the bottom in colder weather, and in summer allow the hot air to escape through the open top. There are no chilling worries.

Both boxes are made of 18mm marine plywood. They can be cleaned and disinfected and will last for years. (Chipboard holds the dirt and gets damp easily. If used, they should be destroyed and replaced each season).

I do not like hollow logs because they are difficult to inspect and impossible to clean properly. They are therefore a potential hazard to the health of your birds. On top of all that they are less convenient to use than nest boxes.

The bottom of my boxes are filled to a depth of 6.0cm - 7.5cm with a mixture of 75% contents of rotten logs and stumps found in the bush and 25% Oregan sawdust. I dampen the mixture with a few drops of water. Neophemas like humidity. During the breeding season I often remove the eggs, add a few drops of water and work it into the mixture with my hand, then replace the eggs. This doesn't seem to worry the hens at all. In the wild nesting hollows the rotted material soaks up moisture from the ground and the hen is constantly turning the mixture over which dries rapidly. I do not use peat any more as I find it too dry.

When I first put the boxes in the aviary I place a square wooden frame in the bottom of the box. This reduces the area from 20cm square to about 14cm square. As the chicks grow or if the clutch is large, or if in hot weather the chicks try to keep away from one another to keep cool, I can remove this frame and give them the extra space they may require.

Hybridising

Hybridising or the breeding together of two different types of Neophema must be avoided. No hybrid Neophemas with a Turk or Scarlet parent have proved fertile to my knowledge. Hybrids from Elegants and Blue-wings have. The only Neophema not to have hybridised with any of the others of the genus is the Bourke Parrot. The Bourke is therefore the safest Neophema to house with others of this family.

Ringing

Whenever I pair up two Neophemas I allocate a specific colour of plastic split ring for their young. I hang these rings above the door of their aviary. When the pairs are changed or the hen is changed, so is the colour of the rings. I keep all these records written down.

When the young are in holding cages and I want to catch an unrelated pair, then all I have to do is catch a pair with different coloured rings. With one-coloured numbered rings, I might have to catch several birds before finding an unrelated pair. This causes unnecessary stress to many birds.

Elegants are rung with split metal rings because they nearly always chew the plastic ones off.

There are many full colour and striped rings available, so many combinations are possible.

If you only have two or three pairs of a particular type of Neophema and you can find the right coloured rings, a useful way of keeping track of "families" of birds at a glance is to ring the cock with say a green ring, the hen with a white ring and all the young with a green & white striped ring.

Keeping Records

Written records must be kept when breeding birds. Records should not only inform you of related birds or what bird is split to what colour. If dates, times, weather conditions, temperatures, box sizes, ages of birds, diets, observations, number of eggs laid, number of eggs hatched, number of young reared etc are noted down each season, valuable research for both your birds and aviculture would be underway. Then personally you could eliminate infertility, poor parents, weak youngsters and generally sub-standard birds from your own birds which would also help aviculture overall. Also answers to why one breeding season is better than another can only be hoped to be answered from correlating relevant information recorded by hundreds of breeders like yourself from all over the country.

Surgical Sexing

Many of my birds are surgically sexed. This procedure not only ensures me definite pairs but also shows up any problems or deformities in their sex organs. From this inspection it is possible to identify potentially infertile birds. Surgical sexing is a great boon to the serious breeder and its use cannot be over emphasised.

In closing this section, I must stress that birds quickly get into a routine. Changes to this routine often upsets them so this should be avoided wherever possible. Good management is the key to success. To rely on Lady Luck is to court with disaster.

Red Pannicum Millet, Mung Beans and Grey-striped Sunflower are soaked in separate containers ▶

▲*A nest of baby Turquoisines*

◀*Turquoisine eggs*

These breeding aviaries are 2.7m long and 2.1m high and 60 cm wide. Brilliant breeding results have been recorded in them from both Neophema & Psephotus Parrots ▶

Bourke Parrot

Neophema bourkii

The Bourke is the only member of the Neophema family not to have green as its basic plumage colour. It is also the only one not to have hybridized with any other Neophema. This trait makes it the most suitable Neophema for sharing an aviary with others of the genus.

This Neophema in its subtle colouring of pinks, blues and browns is well established in aviaries all over the world. Its willingness to breed makes it very popular with beginners.

Bourke Parrots are not very active during the day. However, at dawn and dusk they can be found fluttering and twittering around the aviary. Their larger eyes indicate their preference and suitability for moving around in poor light. This after-dusk activity is the only reason some finch breeders do not put Bourkes in their aviaries, as they can have a disturbing effect on roosting finches.

Generally, these birds are very peaceful, although an eye must be kept open for the odd rogue bird.

Their general management, feeding and breeding has been well covered in the preceding chapter. Usually anything up to seven young can be expected in the nest.

Sexing

The cock bird shows a blue brow, the hen a creamy white brow. As soon as one blue brow feather is seen on a young Bourke you can be sure it is a cock.

Mutations of Bourke Parrot

Cinnamon

This bird is referred to as an Isobella Bourke overseas. It has red eyes and is recessive in its mode of inheritance. Splits are therefore available in cocks and hens. Cocks tend to be darker than the hens and show the familiar blue brow.

Cream

This lovely bird is referred to by some as yellow. The amount of cream markings can vary on some birds. Like the Cinnamon, the Cream has red eyes and is recessive. Cocks show a blue brow as normal, hens do not.

Pink

These striking birds are bred by crossing Rose Bourkes with Cinnamons. There are not too many available at present. The Pink Bourke has red eyes and is sex-linked recessive. Pink Bourkes are sexed in the same way as the Rose Bourkes below.

Rose

This must be one of the most beautiful Neophema mutations available here. The photographs shown should speak louder than words but it is unfortunate that when photographed, the rich salmon pink colour often looks paler on these birds. This mutation is sex-linked recessive. There is a bit of difficulty sexing Rose Bourkes. There is no blue brow on the cocks. Cocks are often a darker colour on the face. They also get increasing grey tones on the face, head and shoulders as they get older.

Pied

Bourkes showing pied type markings in one way or another have been produced. Unfortunately, results in the second generations have been disappointing and this mutation still has to be "fixed".

Red-eyed birds often dislike really bright conditions. Red-eyed Bourke mutations are no exception and shaded or darker aviaries have produced the best results.

▲Bourke cock

◄Bourke hen

◄Pink Bourke hen

Cinnamon Bourke hen (cocks show blue on brow) ►

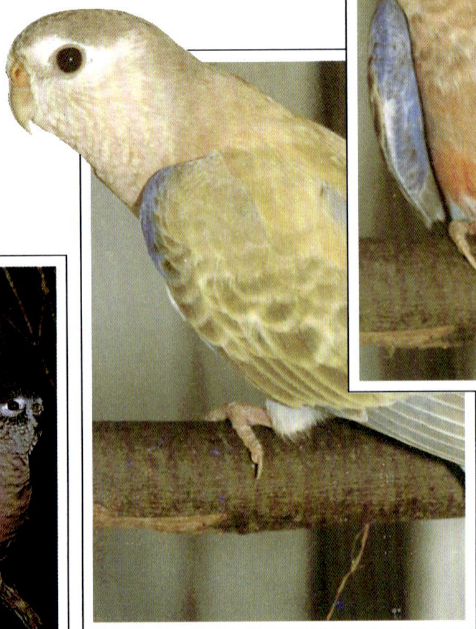

▲Cinnamon Bourke hen

◄Pink Bourke hen and Split to Pink cock

Rose Bourkes are a beautiful salmon-pink ▼

▲ Rose Bourke feeding on Green's tray

▲ Rose Bourke

▲ Cream Bourke

Cream Bourke hen ▶

Cream Bourke hen and Split to Cream cock ▼

Turquoisine Parrot

Neophema pulchella

The Turquoisine is classed as a free breeder and is well established in Australian and overseas aviaries.

This Neophema runs a close second to the Scarlet-chested for colour and beauty. It must be mentioned though that the Turquoisine is more aggressive than other Neophemas. Adult pairs should be housed singly. The only Neophema that should be considered for sharing an aviary with a pair of Turquoisines is the Bourke Parrot. It is preferable that the neighbouring flights do not house other Turquoisines or Scarlets. If they must then a solid opaque partition will avoid bickering through the wire. The swinging perches previously discussed also help when no such partition exists.

Up to six young in the nest is normal but the young should be moved when independent in case they are persecuted by the cock bird, or sometimes the cock bird should be removed.

Sexing

The hen Turquoisine is generally paler than the cock. She has less and not such a rich blue around the face. The blue on her wings is less and paler than the cock. The most obvious difference, however, is that the hen lacks the red/chestnut blaze on the shoulder. (It should be said here that on rare occasions hens have appeared carrying this blaze.) As they colour up, young cocks are brighter than hens and as soon as a red feather is seen coming through on the shoulder it is virtually certain to be a cock.

Another problem often encountered is telling the difference between Turquoisine and Scarlet hens. In these cases it should be found that generally the hen Scarlet is fatter, has more blue on the face, the lores are blue (the lores of the Turquoisine hen are whiter) and the hen Scarlet has sky blue wing coverts where the hen Turquoisine's are a darker blue.

Mutations of Turquoisines

Nearly all aviculturists are familiar with the normal and Red-bellied Turks. There are now a few more varieties being produced in Australian Aviaries.

Full Red-fronted

This is a lovely bird. The red stretches from under the chin right down to the lower abdomen. The green and blue colourings also appear stronger and brighter than the normal. The hens only show red half way up their bellies, not on their chests. For those who like Scarlets but find their aviaries or climate does not suit them, this bird may make an admirable substitute. This mutation is recessive.

Yellow

This striking bird will only become more and more popular in our aviaries. It is a recessive mutation that is well established here in Australia.

Full Red-fronted Yellow

This is obviously a combination of the Full Red-fronted and the Yellow Turquoisine. Red feathering on yellow tends to make the red look more orange. The strength of red on the front of this mutation differs from bird to bird and also in different shades of light. Pairs should be selected for breeding by the strength of red markings they possess. The stronger the red, the better. This mutation is recessive.

Cinnamon

This mutation is still fairly rare. They have red eyes and are sex-linked. When established, they will be combined with other mutations.

Jade and Olives

Like green Budgerigars and green Peachfaces we now have three shades of green in the Turquoisine. The Normal has no dark factors, the Jade has one dark factor and the Olive has two dark factors.

The inheritance of these dark factors can be shown in table form:

0 = No dark factors (Normal Turk)
1 = 1 dark factor (Jade Turk)
2 = 2 dark factors (Olive Turk)

0 x 0 = 100%
0 x 1 = 50% 0 & 50% 1
1 x 1 = 50% 1, 25% 0 & 25% 2
1 x 2 = 50% 1 & 50% 2
0 x 2 = 100% 1

Until the Blue Turk appears or Pied Turks become more numerous, the Jade and Olive Turks cannot be "used" very much. Whether the Olive will dramatically deepen or generally improve the colour of the Yellow Turk remains to be seen.

Pied

These beautiful birds are still rare at the moment. They are a recessive mutation.

Other Mutations

Turquoisines have been bred showing a yellow flash on their wings, rather than a red one. When established, these birds could be used to eliminate the red flash from the Yellow Turk, producing a more yellow bird.

The Blue Turk as far as is known have been occasionally bred overseas but there has been no success producing blues in second and third generations. There are stories about such birds here in Australia but they have not been proved to exist yet.

▼*Jade Turquoisine cock*

▲*Olive Turquoisine cock*

▲*Turquoisine hen*

◀*Turquoisine cock*

▲ *Red-bellied Yellow Turquoisine cock*

▲ *Full Red-fronted Yellow cock and Full Red-fronted split to Yellow hen*

◀ *Pied Turquoisines - cock (below) and his son*

Full Red-fronted Turquoisine cock ▼

Cinnamon Turquoisine hen ▶

Pied Turquoisine (heavily marked) ▶

Scarlet-chested Parrot

Neophema splendida

Other name: Splendid Parrot.

This is undoubtedly the most colourful of the genus. Its beautiful plumage contributes to its world-wide popularity.

This little 20cm gem has broken many beginners' hearts. Many aviculturists describe it as a "soft" or weak bird. It must be emphasised that poor housing with inadequate protection from damp and draught has caused most of the problems for this bird. Scarlets do not like humidity. Better results are obtained in inland aviaries where the days are hot and dry and the nights are cold and dry, rather than on the equable coast. If your birds are uncomfortable in open aviaries then they must be enclosed.

The most obvious difference is that the hen lacks the scarlet chest of the cock bird. She also has less and duller blue colouring on the face.

In young birds the cocks are brighter in colour and any sign of a red feather on the chest will quickly identify the males.

Red-bellied

This is a recessive mutation where a red belly has been added to the cock's red chest producing a Full Red-fronted Scarlet. The hens show only the red belly.

The Sydney Blue

These birds are also called the Sea-greens. They have a green back, cream belly and salmon pink chest. They are recessive. Hens do not have the salmon pink chest.

Par Blue

The back is greeny blue, the belly is creamy white as is the chest. A hint of pink can sometimes be detected on the chest. These too are recessive. Hens do not have the cream chest.

The White-fronted Blue

As its name suggests, this bird is blue with a white chest and belly. It is the most attractive and desirable of the Blue Scarlets. Like the other blues, it is recessive. Hens do not have the white chest.

Cinnamon

This rare bird is sex linked and when established will be extremely useful in producing more mutations. White-fronted Blues and Cinnamons should produce Silver Scarlets for example. Cinnamon cocks have a scarlet chest as normal, hens do not.

Olive Scarlets are established in Australia. Yellow Scarlets are here too. These are yellow with white faces. They are rare and not much is known about them at present.

▲ *Cock Pied Scarlet*

▲ Scarlet-chested cock

▲ Scarlet-chested cock

Scarlet-chested cock ►

Scarlet-chested hen ▼

◄ Scarlet-chested Par-blue hen

▼Scarlet-chested Par-blue hen in nest box

Pair of White-fronted Blue Scarlet-chesteds

Cock Pied Scarlet

Cock White-fronted Blue Scarlet-chested

Pair of Red-bellied Scarlet-chesteds (cock on right)

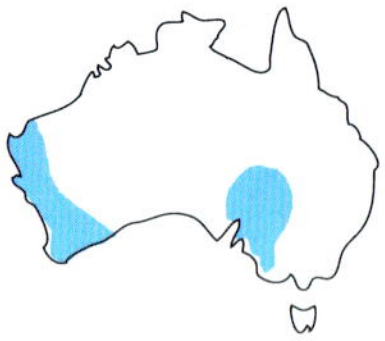

Elegant Parrot

Neophema elegans

This is the first in the group of four Neophemas that are predominantly green in colour. These four, namely the Elegant, Blue-wing, Rock and Orange-bellied Parrot are not as popular as the previous three with the general aviculturist. They are mainly found in the aviaries of specialists (with the exception of the Orange-bellied Parrot).

As mentioned earlier, the Elegant can bomb other Neophemas and is therefore untrustworthy if housed with any others. Cock Elegants are aggressive among themselves. This 22cm Neophema normally has up to six young in the nest and incubation lasts 18 days.

Sexing

Cock Elegants have a more pronounced frontal band than the hen. The lores are a deeper, brighter yellow than those of the hen. The royal blue on the cock's wing travels under the wing and is strong in colour. Under the wing of the hen the blue tends to fade and merge into green. The underwing stripe is usually absent in both mature sexes. If the young however are inspected before they leave the nest, hens often show the white stripe under the wing. Surgical sexing will of course remove any doubt and is to be recommended. How to distinguish them from Blue- wings is described under the Blue-wing.

Mutations of Elegants

Cinnamon

The Cinnamon colour in the Elegants is sex -linked. This is the only well established mutation as yet in the genus.

Lutino

This bird has been bred in Australian aviaries. It is extremely rare at the moment and further information is hard to get.

Blue-winged Parrot

Neophema chrysostoma

Other names: Blue-banded Parrot.

The Blue-winged Parrots are quiet and peaceful and although breeding results often fluctuate from season to season, best results have been obtained through colony breeding rather than as single pairs.

With the exception of the Orange-bellied Parrots of which little is known in aviculture, the Blue-winged Parrot is the poorest breeder of the genus. Breeding in the wild in Tasmania, S-E South Australia, Southern Victoria as well as in aviaries in Northern Europe indicates this bird's dislike for hot climates. Over-heating should therefore be watched with this species.

Sexing

The blue frontal band is less pronounced in the hen. The lores of the hen are not as bright or yellow as with the cock. The blue on the wing is duller than the cock and the blue on the underwing coverts in the hen is duller and quickly merges into olive green. Young hens tend to show the underwing stripe before they leave the nest. Distinguishing features between the Blue-wing and the Elegant are:- The Elegant has a double frontal band, the lower band being dark blue and the upper band is pale blue. The Blue-wing has a more uniform dark blue band. The Elegant's frontal band extends above and behind the eye, the Blue-wings does not. The Elegant tends to be a brighter more yellow colour especially on the chest, the Blue-wing tends to be greener. The Elegant has less blue on the wing than the Blue-wing.

Mutations

None.

Some Elegants have a ▲ red "thumb mark" on the lower abdomen.

Cinnamon Elegant cock and normal hen ▲

▼Elegant hen

▼Elegant cock

◀ *Blue-winged hen*

▼*Blue-winged cock*

Distinguishing Elegants from Blue-wings

The most distinguishing points are illustrated below

The Elegant has a double frontal band, lower band being dark blue, upper band is pale blue. Also, the band extends above and behind the eye.

Elegants tend to be a brighter yellow on the chest.

The frontal band on the Blue-wing is more uniform dark blue and does not extend behind the eye.

The Blue-wing is generally greener and has more blue on the wing than the Elegant

Rock Parrot

Neophema petrophila

The Rock Parrot is only found in aviaries in Australia today. It is now unknown in aviculture overseas.

Many people find this bird difficult to breed. There are two points that must be stressed here. The first is that Rock Parrots do not eat a lot of seed. They should be mainly fed on vegetables, fruit, seeding grasses, arrowroot biscuit etc as described in general feeding. In the wild they eat a lot of Pig-face, a wild succulent that grows on beaches. If they have no choice they will eat only seed and as a result tend to become overweight. This will reduce their breeding potential and their lifespan.

The second point is that they have a different breeding habit to other Neophemas. With other Neophemas you will find that if the young leave the nest slightly early and you put them back in the nest box, within a short period of time they will be out again. You just cannot get them to stay in the box. The young Rock Parrots, however, continuously go back to the nest box and often camp in the boxes at night, even when the hen is incubating a second clutch. They are very much a community bird and should be kept as such. They are active and enjoy playing with and chasing each other. Best breeding results are obtained when these birds are housed in colonies. Three pair colonies have proved extremely successful.

The Rock Parrots will use normal nest boxes. There is no need to provide them with the rock crevices they use in the wild.

Sexing

The dark blue frontal band is narrower on the hen than the cock; otherwise they are similar. Surgical sexing is a must for this species.

Mutations of Rock Parrot

Yellow Rock Parrots have been bred. No further details are known at present.

Rock Parrots will breed in nest boxes. Rock crevices are not required in aviaries

Rock Parrots are hard to sex visually

Orange-bellied Parrot

Neophema chrysogaster

The Orange-bellied Parrot in appearance is a typical member of the Genus Neophema with many characteristics common to most others of the group. It is a small, slender, basically green parrot of approximately 20 cm length and weighing about 55 grams. In common with other Neophemas it is a ground feeding seed-eater, but unlike the Ground Parrot (Pezoporus wallicus) whose habitat it shares during the breeding season, the Orange-bellied Parrot rests and roosts above ground, and in a typical parrot manner, nests in hollows and spouts of Eucalyptus trees.

It is the most temperate of the genus, breeding in far South Western Tasmania and winters, after the leisurely northwards migration, along the coastal fringes of Victoria and South Australia.

▲ *Cock Orange-bellied Parrot*

Almost unknown in captivity, past efforts to maintain and breed the Orange-bellied Parrot have met with very little success and prior to 1986 the only successful captive breeding in Australia was recorded by the late Fred Lewitztea of Adelaide in 1973 and 1974 when two and one young were reared to independence. Numerous other attempts have always ended in failure. Outside Australia, it is understood that limited success was achieved in Europe in the early 1970's when young were reared in Holland and West Germany but they appear to have died out there. Concerns for the plight of the species in the wild led to an extensive survey which commenced in 1979 and which resulted in the conclusion that the total species population numbered fewer than 150 birds remaining in the wild.

With so low a total population it was essential that immediate action should be taken if the species was to be saved from extinction. Management initiatives were undertaken to protect and improve the habitat available to the birds both in their breeding range in South West Tasmania and on the mainland, especially in Port Phillip Bay in Victoria where important over-wintering habitat occurs.

The Orange-bellied Recovery Team of Scientists from the Commonwealth and States of Victoria, Tasmania and South Australia and with representatives of the Royal Australian Ornithologists Union and the International Council for Bird Preservation, under the Chairmanship of J Forshaw, recommended in 1985 that a captive breeding programme be established and be carried out by the Tasmanian National Parks and Wildlife Service.

The programme commenced in 1986. In the first year (1986/87) both of the two females held went to nest and four young (3 and 1) were reared to independence. In the second year, three of four females held nested and eight young (4, 3 and 1) were reared to independence. The pair which reared one young let a further two young die in the nest. In 1988/89, all seven females being held nested and twenty two young were reared to independence.

The primary object of the programme is to repopulate the wild and it is hoped that it will be possible to make the first releases back to the wild in 1990.

Prior to taking Orange-bellied Parrots into captivity a trial programme was undertaken using two other Neophema species, the Rock Parrot (N. petrophila) and the Blue-winged Parrot (N. chrysostoma). Both these species bred in abundance in the aviary complex. 86 individuals of the two species have been released back into the wild from 1985-88.

A number of lessons have been learned in this programme so far in breeding these three species which are probably applicable to all Neophemas and will help with future breeding programmes.

1. Natural selection of pairs by allowing several pairs to be run together: Both the Rock Parrots and Orange-bellied Parrots have proved lethargic in captivity in past efforts to breed them. By running several pairs together, birds are kept active vying with each other for food, mates, etc and as a consequence they remain fit and healthy. In our conditions infertile eggs are a rarity.

In this programme the aviaries are basically 7m x 4m x 2.5m, high with two ranges of five aviaries side by side. The partitions can be removed to double or treble the aviary size. All the aviaries were either double (8 x 7m) or treble (12 x 7m) for the breeding programme. In each case between two and five pairs were run in each aviary. The only problems associated with this breeding method is that when the nest boxes are first put into the aviaries, so great has been the exuberance of males in encouraging the females to nest that they are liable to drop dead off the perch. In 1985, five male Rock Parrots died in this manner and in 1988, two male Orange-bellied Parrots died on the same day. Stress brought on by the excitement was the cause.

In both 1987 and 1988, there was an excess of one female over male Orange-bellied Parrots in three of the aviaries. On each occasion one male must have mated with two females for all females in the aviaries laid fertile eggs. This is unusual with monogamous birds but it is quite probable that a similar occurrence will take place with all Neophemas under the same circumstances.

2. Natural nesting logs: There has been limited use of artificial nest boxes in this programme, but natural logs have proved more successful than manufactured boxes. In general, small diameter logs are better than large diameter logs. Females respond best to logs which have a base area into which her body fits snugly - little more than 10cm in diameter is ideal. There was no clear evidence for a preference for either vertical or angled nest logs as both were used equally well.

Nest boxes were erected in the aviaries in mid August. Perhaps if they were put in a little earlier, the mortality which occurs amongst males would not occur.

During winter most birds are run together as a group similar to their behaviour in the wild. Most Australian Parrots are colonial outside the breeding season and in captivity they should be allowed to behave as naturally as possible.

All the Neophemas are fed on basic Canary mixture with added Linseed, Niger, Japanese Millet and Rape Seed. Shellgrit and cuttlefish are available at all times. During mid to late winter (July-September) S B 40 conditioner is added to the seed twice per week and, when the young hatch, seeding milk thistle, plantain, dandelion grasses, etc are fed every second day.

The aviary base is sterile Blue-metal to prevent the cycling of worm infestations.

In 1988/89 an attempt was made to cross foster six Orange-bellied Parrot eggs under Rock Parrots in two nests. All eggs failed to hatch. Four were destroyed by the Rock Parrots and the other two failed to hatch. Further efforts will be made to cross foster eggs to try to obtain double clutches in the future. It has worked with other species and there is no obvious reason why it should not be possible with Orange-bellied Parrots.

In summary, most Neophemas are ready breeders in captivity. Through our experience we have found that the Orange-bellied Parrot is no more difficult than any others of the genus.

It is most important that both the captive breeding programme and the wild management programme should succeed in order to save this magnificent little Parrot for the future.

▲ *Hen Orange-bellied Parrot at nesting site*

Psephotus Grass Parrots

Introduction

The Psephotus Grass Parrots can be broken down into three groups:-

1. **Red-rumped** (Psephotus haematonotus) and Mulga Parrots (Psephotus varius)
2. **Hooded** (Psephotus chrysopterygius dissimilis), **Golden-shouldered** (Psephotus chrysopterygius chrysopterygius) and **Paradise Parrots** (Psephotus pulcherrimus)
3. **Red-vented** (Psephotus haematogaster haematorrhous), **Yellow- vented** (Psephotus haematogaster haematogaster) and **Naretha Blue- bonnets** (Psephotus haematogaster narethae)

Like the Neophemas many of this group are becoming more and more popular. Some of them, like the Red-rumped and Hooded Parrots, also have their mutations. Although not large birds, all members of the group are aggressive and especially spiteful to their own kind. An aviary unto themselves is a definite rule for Psephotus Grass Parrots.

The Hoodeds, Golden-shoulders and, if anyone has them, the Paradise Parrots must rank among the most beautiful of parrots available to Australian aviculturists today. The Blue-bonnet is more colourful than many realise. The pretty Mulga with splashes of yellow and red on green brings us to the Red-rump, maybe not as brightly attired as the rest, but certainly full of the one common factor that binds this group together, and that is character. When you own some Psephotus Grass Parrots you will soon learn that they are all individuals - nosy, cheeky, aggressive in a nice sort of way and arrogant. For this trait alone they make their presence in a collection worthwhile.

Management

Points made in the management and feeding of Neophemas also apply to the Psephotus group, the obvious exception being the swapping around of cocks and hens in the breeding season. This practice would possibly have fatal results with the Psephotus.

Housing

Let it be repeated - one pair per aviary. No other birds should share their accommodation. Ideally, solid partitions should divide the flights as this will avoid bickering and fighting between neighbours. (Double wire is the minimum division that should be considered). Being able to hear but not see others of the genus most certainly helps cocks to come into breeding condition - ready to defend his mate and his nest site. Aviaries between 75cm and 1m wide, 2 - 3m long and at least 2m high will adequately cater for the needs of these birds. A solid shelter at least 1m deep similar to shelters mentioned under Neophemas will provide shade and a weather proof retreat. Many breeders now completely cover the roof of the flight with a suitable transparent material. This eliminates all weather problems, disease problems that can be passed through a wire roof by wild birds, and helps reduce cat and hawk troubles. Ample sunshine will penetrate the aviary through the front wire, providing of course the aviaries are facing north to north east. The advantages of concrete floors are well known - easy to clean, helps control worm infestations, keeps out mice etc. If dirt floors are to be used, make sure tin or concrete walls are sunk at least 45cm into the ground all around the aviary to prevent vermin from digging in. Cover the roof completely with clear fibre-glass or similar - dirt floors are one thing, muddy floors are quite another. To reduce problems with fallen seed, food scraps and worm eggs, dirt floors must be kept as dry as possible. Paving stones are also used successfully on aviary floors.

It should go without saying that a safety corridor or individual safety doors must be used. It is really inexcusable, if you think about it, to let a bird escape from your aviary.

Each aviary should contain two main natural perches (Eucalypts etc), one in the shelter and one near the end of the flight. An aviary full of perches can be hazardous to the birds and owner and also encourages the birds to hop from one end of the aviary to the other rather than fly. Birds get maximum exercise by flying up and down. It is therefore a good idea to have food and water receptacles below the perch level (not directly under of course). This means the bird has to fly down to feed and drink and back up again to the perch.

Feeding

The feeding section under Neophemas is quite sufficient for the Psephotus group. There is not much more to add except that under the specific bird section that follows, various other foodstuffs such as madeira cake and peas will be mentioned. Your birds should be offered all such items to see what they will and will not eat. Vitamins and minerals in the drinking water are positively no substitute for a good varied diet.

Breeding

Bobbing up and down, strutting and hovering around the hens, feeding each other etc. are all signs of courtship displays. Individual breeding requirements are discussed under the relevant bird.

All in all, it will be found that Red-rumps, Mulgas and Bluebonnets have basically the same management and breeding requirements. Hoodeds and Golden-shoulders are both similar in their requirements as well.

▲ Pair of Red-rumped Parrots

▲ Pair of Cinnamon Red-rumps

Red-rumped Parrot

Psephotus haematonotus

Other Names: Grass Parrot, Grassie

These charming character birds must contend with the Eastern Rosella and Cockatiel as the ideal beginner's parrot. It is a pity that out of the three the Red-rump and Eastern are aggressive birds. Many beginners have only one aviary and are tempted to mix varieties of birds together. Invariably problems arise between these two species and other birds. It therefore must be stressed that they should be housed in single pairs. Hence when you are starting out in keeping parrots, two aviaries 3m long and 1m wide is a far better proposition than one aviary 3m long and 2m wide. Red-rumps will do well in aviaries 75cm to 1m wide, 2m to 3m long and 2m plus high. The general housing rules mentioned previously of course apply.

▲ *Pair of Lutino Red-rumps (Cock on left)*

▼ *Pair of Lutino Red-rumps*

Hen Cinnamon Red-rump ►

◄ *Cock Cinnamon Red-rump*

Feeding

Red-rumps should present no problems in this area to their owners. They usually eat everything given to them.

One successful Sydney breeder feeds a mixture of Japanese, French White, Pannicum Millet and Canary Seed and occasional green feed during the non-breeding season. None to very little Sunflower seed is given. During the breeding season the birds get Sunflower, sprouted mung beans, greens, seeding grasses and a 2cm block of Madeira cake every day as well as, of course, the staple mix of Canary and Millet seeds. The birds have no problems due to being over fat and the increase in foodstuffs encourages them to begin breeding. He uses the same technique with great success with his Mulgas.

Breeding

The recognised breeding season is from August to December, but like many of our aviary bred parrots, if it suits them they will breed. It would therefore be safe to say that they could breed any time between May and January.

Logs or boxes can be used and should be about 45cm deep and 15cm square. Wire or timber ladders can be attached inside the boxes. Logs are often better hung just off the vertical to reduce the chances of the parents jumping straight down onto the eggs.

A compatible pair of Red-rumps is a key to success. These are always sitting and feeding together, preening one another etc. It is not uncommon for them to have four broods in a season and two broods are pretty well normal. The cock bird feeds the hen while she broods alone. Four to seven eggs (usually five) are laid and incubation takes about twenty days. The young nearly always hatch on the same day. Certainly all are hatched over a 48 hour period. A further four to five weeks and the young will be leaving the nest. They are usually independent two weeks after. Young birds can be left longer with the parents but the parents must be watched for any signs of aggression. A careful eye must be kept on the hen. She often resents the young (especially young hens) when she begins her second brood. The cock usually presents no problems.

The young can be sexed in the nest. The young cocks are brighter in colour and red feathers can be seen coming through on their rumps.

Red-rumps will breed between 12 and 18 months of age. Cocks tend to mature quicker than hens. All in all the Red-rump is a no- problem bird and very willing to breed.

Mutations of Red-Rumped Parrot

Cinnamon

Genetically, there are two Cinnamon varieties in the Red-rumped Parrot. One is sex-linked, the other is recessive. They cannot easily be distinguished visually. Recessive Cinnamons sometimes show a creamier coloured beak and lighter coloured legs than the sex-linked variety. Cinnamon Red-rumps do vary in colour tones between themselves. Some are a dull biscuit colour, others lean towards a bright lemon shade. The more prevalent Cinnamon appears to be the sex-linked variety. Cock Cinnamons have a red rump, hens do not.

Lutino

This is another sex-linked mutation. The cock is bright yellow with a few paler to white feathers on the wings and tail. He still retains the red rump which looks very striking on yellow. The hen is paler than the cock. Like all ino's they have red eyes.

Black-eyed Yellow Red-rump

This is still a rare variety in Australia. Not as bright as the Lutino, it is yellow and has normal black eyes. This is yet another sex-linked variety.

Blue Red-rump

Not as rare as it used to be, but still far from common. The beautiful Blue Red-rump is going to play a major role in the development of future Red-rump mutations. The Blue Red-rump is recessive. Both cocks and hens can therefore be split for blue.

Sexing: The cock bird has a blue head and darker blue shoulders, a white rump and blue tail. The rump in the hen is blue, there is no white at all.

In the future, Blue should combine with Lutino to produce Albino, with Cinnamon to produce Silver, with Black-eyed Yellow to produce Whites and with Pieds to produce Blue Pieds - it's all ahead of us at the moment.

Pied Red-rump

This mutation is rare. It is recessive. I'm sure it would look terrific in blue and white too - one day.

Opaline Red-rump

This mutation is sex-linked. The photo does give us some idea of how beautiful it is. There is a definite resemblance to an Opaline Budgerigar.

▲ *Baby Red-rumps about three days old*

▲Blue Red-rumped hen

Yellow Red-rumped hen (Black-eyed) ▶

▼Blue Red-rumped cock

Pied Red-rump ▶

Back view of Pied Red-rump ▶

▲Black-eyed Yellow Red-rump hen and Split Cock

▼Opaline Red-rump

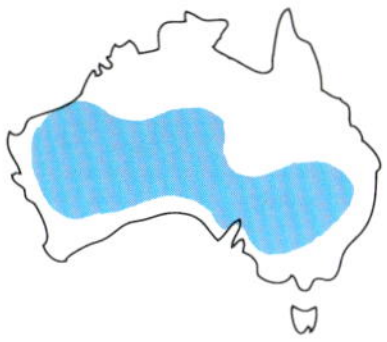

Mulga Parrot

Psephotus varius

Other names: Many-coloured Parrot, Varied Parrot.

The Mulga or Many-coloured Parrot's feeding and housing requirements are more or less the same as the Red-rumps so I will not repeat what has already been mentioned. Less common in aviaries than the Red-rump, the Mulga is not as aggressive, but one pair per aviary still produces the best results.

▲ *Pair of Mulga Parrots (cock on right)*

Mulgas can be double brooded but they are not as prolific as the Red-rump. The hen alone broods, incubation lasts about 20 days and all young hatch within a 48 hour period. On fledging, the babies are much more mature looking than the Red-rumps. They have more colour, their tails are longer and they are less flighty. They do stay in the nest about a week longer than the Red-rumps. Young Mulgas can be sexed in the nest. Cocks show yellow feathers on the shoulder and hens red. When the young are being fed it will be noticed that the larger food particles such as Sunflower, sprouted mung beans, Madeira cake etc. disappear very quickly. This is because it is easier, quicker and therefore less tiring for the parents to fill their babies' crops with these foodstuffs than collecting small millets and the like.

Mutations of Mulga Parrot

Cinnamon

Recently, what looks to be a Cinnamon Mulga has been bred. No more details are available at this point in time. Hybrids between the Lutino Red-rumps and Mulgas should be watched out for and avoided.

▲ *Cock Mulga Parrot*

Blue-bonnet

Psephotus haematogaster

There are three Blue-bonnets: The Yellow-vented Blue-bonnet (P. h. haematogaster) - Red abdomen with yellow vent. The Red-vented Blue-bonnet (P. h. haematorrhous) - Red abdomen with red vent. The Naretha Blue-bonnet (P. h. narethae) - Yellow abdomen with red vent.

The Naretha Blue-bonnet is much smaller and rarer in aviaries than the red or yellow venteds.

The Blue-bonnets are one of the most misunderstood birds available to aviculture today. They are regarded as noisy, drab, extremely aggressive and poor breeders. From the previous sentence, only the aggressive tag really fits. Most of our information is passed down over quite a considerable time and little or no up dating due to more advanced aviary management is available.

The Blue-bonnets certainly are aggressive birds but probably no more so than the other members of the Psephotus group. Gram for gram, the Hooded Parrot is by far the most aggressive member. Although the Blue-bonnets have some raucous calls, I doubt if it could be called noisy when compared to Lorikeets for example. Although from a distance Blue-bonnets appear mainly brown, on closer inspection they are very pretty with splashes of blue, red and yellow.

Blue-bonnets if housed sensibly and given a suitable nesting receptacle should and usually do provide a rewarding experience. Blue-bonnets must be housed as single pairs and mixed with no other species of birds. Double wire or solid wall partitions are essential. Aviaries should be completely roofed as should all aviaries. An aviculturist's greatest enemy is the weather. Birds do not need to be rained on and if your aviaries face in a northerly direction (as they should) ample sunlight will be provided. If you want to wet your birds, install a sprinkler system. At least one third of the aviary roof should be steel, aluminium or fibro to provide shade and shelter. The remainder of the flight should be covered with clear fibreglass sheeting or similar.

A simple seed mixture of mixed millets and canary seed with a little Sunflower seed is all that is necessary as a basic diet. Green seeding grasses should be supplied whenever available and also fresh corn on the cob, apple, spinnach and soaked and sprouted seeds. During the breeding season all of the above foods should be increased and plain Madeira cake supplied, especially while chicks are in the nest.

Blue-bonnets nest early in the season, usually late July. The birds are fairly secretive about their nests and require a log of fairly strict dimensions before they will accept it. A true pair should show interest in a log of about 60cm with an internal diameter of 15cm. The entrance hole should either be a small natural knot hole or a hole cut about 10cm down from the top of the log and measuring about 5 cm in diameter. Logs should be hung almost

▲ *Pair of Red Vented Blue-bonnets*

This is why it is called a Red-vented Blue-bonnet. The Yellow- vented has the same red abdomen but a yellow vent

Wide-eyed with head feathers raised, a Red-vented Blue-bonnet struts along the perch

Frontal view of Naretha (left) and Red-vented Blue-bonnets

Back view of Naretha and Red-vented Blue-bonnets. Note the size difference

upright and filled to a depth of about 7.5cm with decayed wood pulp, crushed termite nest or similar. The hen usually sits very tightly and disappears into the log on sighting anyone. Nest boxes are accepted by some pairs but a suitable log usually has more appeal. Nest boxes should be about 60 cm high and 20 cm square.

The usual clutch is about five eggs but some pairs will lay up to seven. The hen broods and incubation lasts about 20 days. The young remain in the nest for about five weeks. About two weeks after fledging the young should be independent. A watch must be kept on the parents for any signs of aggression towards the young.

For obvious reasons, Red and Yellow vented Blue-bonnets should not be interbred.

Mutations of Blue-bonnet

White Blue-bonnets do exist in Australian aviaries, however, further details are not available just now.

The Hooded & Golden-shouldered Parrots

Psephotus chrysopterygius

Golden-shouldered Parrot (P.chrysopterygius)
Hooded Parrot (P.dissimilis)

These two beautiful birds are very similar in all aspects and will therefore be discussed in one chapter, thus avoiding much repetition.

Gram for gram, the Hooded Parrot is probably the most aggressive bird discussed in this book. If compatible pairs cannot be purchased, then unrelated, uncoloured birds should be selected. Hopefully these will have paired up when fully coloured at around 12 - 16 months of age. It is often difficult to find compatible partners for single mature Hoodeds.

The Golden-shouldered Parrot is not as aggressive but the same method of matching pairs is recommended.

These two birds are growing in popularity every day. The number of people who have problems with these birds is quite alarming. For these birds, the old saying that "there are no hard and fast rules in aviculture" must be completely ignored. Unless the strict basic rules are adhered to, success with these birds will be very limited.

Housing

Firstly, there should be only one pair of Hoodeds or Golden-shoulders per aviary. No other birds should be kept in the same enclosure. An ideal size is 3m long by 1m wide and 2m high. Any longer is a waste of time, material and effort. Longer flights allow such swift flying birds time to build up momentum which can prove fatal to young birds or night flyers on suddenly contacting the end wire.

A solid shelter must make up at least one third of the total aviary length to provide shade and protection. The remainder of the flight should be completely roofed in with a semi transparent or transparent material. Solid or at least double-wired partitions should divide each aviary containing these birds or anything else. Concrete floors should be covered to a reasonable depth with river sand or the like. These birds love to dig. Other points mentioned in this book, such as direction faced and points on cleanliness, draught and damp etc, of course apply.

Feeding

Many food items already mentioned will be adequate for Hoodeds and Golden-shoulders. Seed mixtures should be kept simple eg. Canary, Pannicum, Red and Japanese Millets are mixed in equal proportions. Grey-striped Sunflower and Hulled Oats are given at about half the amount of the small seed mix. Green feed is greatly appreciated especially green peas. Seeding grasses should be given as available. A successful Sydney breeder soaks seed, Canary and Millets only, overnight, in a mixture of half a teaspoon of honey and 1 teaspoon of Calcium Sandoz

◀ *Hooded hen*

Cock Yellow Hooded and split to Yellow hen ▼

Hooded cock ▶

▼ *Pair of Hoodeds*

◀ Golden-shouldered hen

Pair of Golden-shoulders ▼

▼ Golden-shouldered cock

Golden-shouldered cock ▼

Syrup to half a litre of water. The amount of seed used would fill a 2 litre container. All moisture is absorbed by the seed. The seed is then rinsed and fed to the birds. This is done once a week. Every second week six drops of Cod Liver and Wheatgerm Oil is added. Calcium is necessary for the well being of adults and chicks and these birds seem to need a copious supply. Cuttlefish bone and grit should always be available. If the Cuttlefish is not being touched replace it, reposition it or break it up.

Breeding

In aviaries these birds will usually breed between February and September. Courtship displays in both birds are similar. The cocks raise their frontal head feathers to form small crests. They puff their chests out and strut and flutter around their hens.

Four to six eggs are laid and incubation lasts about twenty days. Only the hen broods. She is fed by the cock. The young leave the nest at about five weeks of age and they are independent anywhere between 10 and 20 days afterwards. The young should be removed as early as possible - as soon as they are seen eating by themselves from the seed dishes. If you leave the young in the aviary they will winge for four weeks or more. This will not only impede the parents from nesting again but it will also tire them by their youngsters' continual demand for food. Aggressive parents, especially the cocks, must also be carefully watched for. Young cock Hoodeds and Golden-shoulders can be identified by their brighter blue cheek patches. They will be fully coloured up by 18 months of age, but cocks have successfully bred while still colouring up.

Although they nest in termite mounds in the wild, in the aviary nest boxes are sufficient. Heated nest boxes are not necessary but the nest chamber should be just big enough for the hen and her babies and be insulated. What has proved a very successful nest box has an outside box measurement of 30cm by 20cm by 20cm. The internal nest chamber measures 15cm by 15cm by 15cm without any ventilation holes - the entrance tunnel or spout provides adequate air. The gap between the two boxes is insulated with polystyrene or similar. Each box has its own separate lid and a spout 6 ins long passes through the first box into the inner chamber. PVC piping, grooved inside to provide grip for the birds, is adequate for the spout. The spout should be just big enough for the birds to squeeze through, about 6cm in diameter.

These birds come from areas in which the temperature is seldom below 30C at any time of the year, and their tunnels in termite mounds are the only method of air circulation to the nesting chamber. It is for this reason that wild Hoodeds and Golden- shoulders can leave their nests for long periods of time. The heat from the live termite mounds is about 38C and this temperature keeps the eggs safe. Finely crushed termite mound (tree type) or decayed wood pulp is used as nesting material. This is soaked in water for a few days before being squeezed and placed damp in each nest box to a depth of about 5 cm. Do not be alarmed if most of the material is kicked out. The nest boxes should be cleaned out and material replaced after each nest. The boxes can be placed at any height from 1m off the ground to the roof. The birds seem more at ease if the boxes are facing the front of the aviary as they can watch for any approaching danger.

The birds should be wormed at least twice a year. Birds should not be wormed during the moult or all growing flight and tail feathers will often fall out. There appears to be no permanent damage done, as the next feather growth is usually normal.

In conclusion, check your birds morning and night and never be afraid to check nest boxes. Enter your aviaries regularly and earn the trust of your birds. If you want them to be good to you - be good to them. If a bird becomes ill, treat it immediately or take it to an Avian Vet - don't wait to see if it will improve. Keep detailed records of all happenings in your aviaries as they will prove invaluable in later years.

Little more can be said except that these birds are not generally easy to manage (again there will be exceptions) and unless you give what they need, not what you want, you will never be a successful breeder of Hooded or Golden-shouldered Parrots.

Mutations of Hooded & Golden-shouldered Parrots

Yellow

What a colour! This has to be one of the deepest, richest yellows seen on any feathered mutation. The Yellow Hooded is a recessive mutation that must simply grow in popularity as more are bred and more people see this very striking bird.

Pied

Yes - another Hooded mutation. Pied Hoodeds are here in Australia but no details are available at present.

There are no known mutations in Golden-shoulders at the moment.

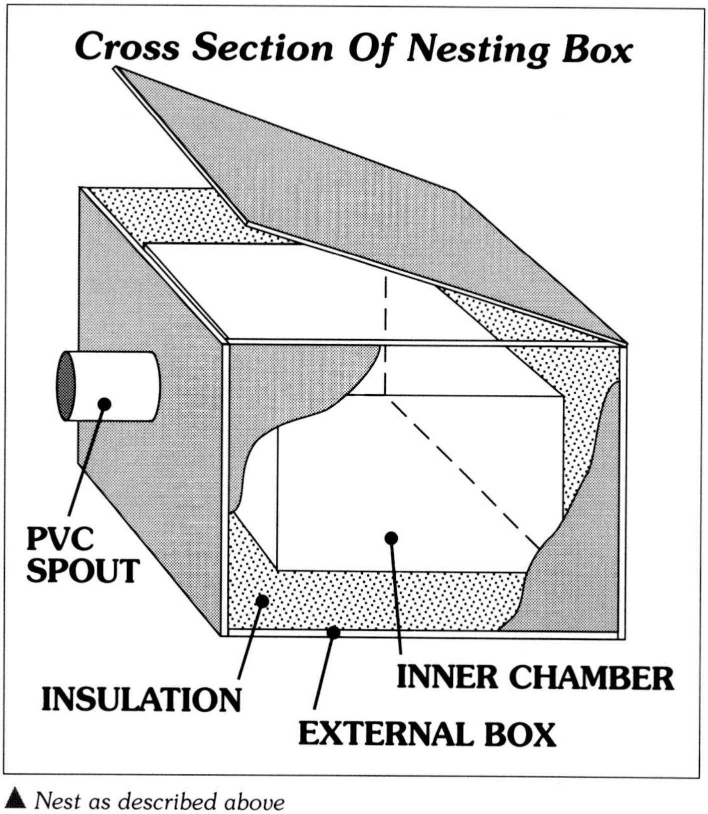

▲ *Nest as described above*

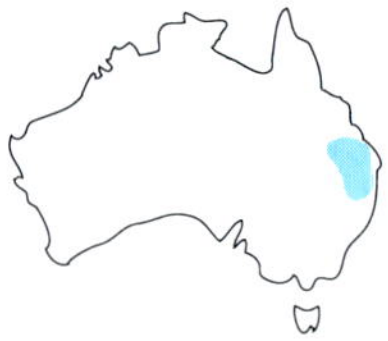

Paradise Parrot

Psephotus pulcherrimus

Last confirmed sighting 14th November 1927!

It is extremely doubtful that this gorgeous bird exists in captivity today. Those that have been shown as Paradise Parrots have all been hybrids, usually between Golden-shoulders and Mulgas.

Three views of Cock Paradise Parrot

From Left to Right: Cock Hooded, Cock Paradise, Cock Golden- shoulder. Note the difference in size between the cock Hooded and the cock Golden-shoulder.

Pair of Paradise Parrots

Hybrids like this are the first steps in trying to create a "man- made" Paradise Parrot. This bird is showing the brown back, red wing flashes and greeny-blue front. Experimenting like this should not be encouraged.

Pair of Golden- shouldered Parrots

Pair of Hooded Parrots

Periodically there are odd reports of sightings in the wild. Some of these can sound quite reasonable. A confirmed sighting is bound to lead to three immediate responses: 1. Simply hundreds of people flocking to the area hoping for a glimpse. 2. Illegal trapping. 3. Legal trapping by National Parks and Wildlife Service in the hope that various experts can breed and save these birds.

In view of this, if someone is out there now looking at a Paradise Parrot and has no intention of telling anyone about it, then no-one can blame him.

The name Paradise Parrot will remain in the Red Book. Entries in this book are those that have not been seen for at least fifty years. Should Paradise Parrots be made officially extinct and some turn up, they will not be protected under the law. There are various factors that contributed to the decline (or extinction) of the Paradise Parrot. The main one would have to be habitat alteration. This bird is, or was, fairly dependent on grass seeds. The Aborigines used to burn the land quite regularly. The Europeans, however, burned the land and then ran cattle. This reduced grass growth drastically. Added to the fire and grazing would be introduced and feral animals (cats, rats, foxes etc) and pressure from trapping mainly for the overseas market.

There is nothing to support the theory that Paradise Parrots are a natural hybrid. The Hooded, Golden-shoulder and Paradise Parrots are all similar and closely related birds. Usually if there is an overlap of territories of such birds in the wild, there is a reinforcement of what are called isolating mechanisms. This makes the birds genetically incompatible. Hybrids between them are infertile or at the extreme cannot occur. Their territories, however, did not overlap so hybrid problems did not exist. Golden-shoulders and Mulga Parrots would also never meet up in the wild.

Many Neophemas' ranges overlap and this is why they are genetically incompatible.

Aviary hybrids do not come into the reckoning because events can take place in aviaries that could not happen in the wild. Expeditions to find the Paradise Parrot are organised periodically but quite often they don't even find a hole in the termite mound.

If aviculture, as we know it today, had existed in the early 1900's then these birds would probably still be with us.

We can only hope that someone, somewhere is watching these exquisite birds.

Diseases of Neophema & Psephotus Grass Parrots

An Introduction

Many of the disease problems encountered in these birds (eg. worms, mites & lice, bacterial and some viral diseases) can be prevented by proper management. The Poultry industry learned many years ago what aviculturists are finding today: proper quarantine procedures, aviary design, pest control, good diet and cleanliness are essential to maintaining a healthy collection. Proper management prevents disease entering the aviary and controls disease within the aviary.

There are several topics which will be discussed under the following headings:

1. **The Health Check of the newly acquired bird**
2. **Quarantine of the new bird**
3. **Stress and disease**
4. **Recognising disease**
5. **Emergency procedures in the event of disease**
6. **First aid for the ill bird**
7. **How to get the best out of your avian veterinarian**
8. **Control and prevention of disease in the aviary**
9. **Worms and a worming programme including average weights of the birds**
10. **The best method of dosing and administration of drugs**
11. **How and when to use the crop needle**
12. **The proper use of drugs**
13. **The more common diseases found**
14. **Notes on some diseases of baby birds**
15. **Some reference points for hand rearing**
16. **Surgical sexing**

1. The health check of the newly acquired bird

The newly acquired bird represents the most common disease danger to the aviary because some diseases are difficult to detect. It is necessary to properly examine any new bird before it enters your aviary. This examination starts at the head, checks the eyes, the nostrils, the feathers around the nostrils, the head, body, wings and especially the tail for signs of abnormal feathers, the vent for signs of pasting and the keel bone for "going light". The birds must be checked carefully for watery eyes and discoloured or malformed nostrils. These are signs of conjunctivitis and sinusitis. These diseases are especially prevalent in the Bourke, Turquoisine, Elegant, Blue-bonnet and Red-rumped Parrots. Left untreated for any period of time, these diseases often become impossible to cure. Both are symptoms of Chlamydiosis. This is a particularly dangerous disease to birds as well as humans and you must be able to recognise its symptoms. Birds which have a prominent keel bone are said to be "going light". This has many causes, most of which are dangerous to your resident flock. Such a bird should not be purchased. It is very important to check the feathers. Many feather abnormalities seen in this group of birds indicate virus diseases such as Papovavirus or Feather or Beak Virus. These diseases are usually not detected until the breeding season. It is advised not to purchase Hoodeds and Golden-shoulders during a moult.

The health check of the new bird is the first step in preventing disease from entering your aviaries. The quarantine procedure is necessary to detect any diseased birds which passed the new bird health check. Some birds appear healthy but in fact may carry a disease. These birds are called "carriers" and quarantine procedure allows us to detect such birds. Many of these diseases are stress related and within the six week quarantine period most of these diseases will become apparent.

2. Quarantine of the new bird

Quarantine gives the birds time to acclimatise to their new surroundings, so that by the time they enter the aviary they are not in a stress situation. This lessens the chance of introducing disease to your aviary. During the quarantine time Neophemas and Psephotus Grass Parrots should be treated routinely for worms and mites using Ivermectin. (This is a restricted drug and is available only by prescription). The advantage in using Ivermectin is that it will kill both the worms and the mites in this group of parrots. The only birds that can be housed together in the quarantine area at one time are those of a compatible species and from the same breeder. Any bird showing signs of illness must be immediately isolated. All healthy quarantined birds are allowed to enter the main aviary after six weeks. The quarantine aviary is then disinfected and cleaned in preparation for the next new purchase.

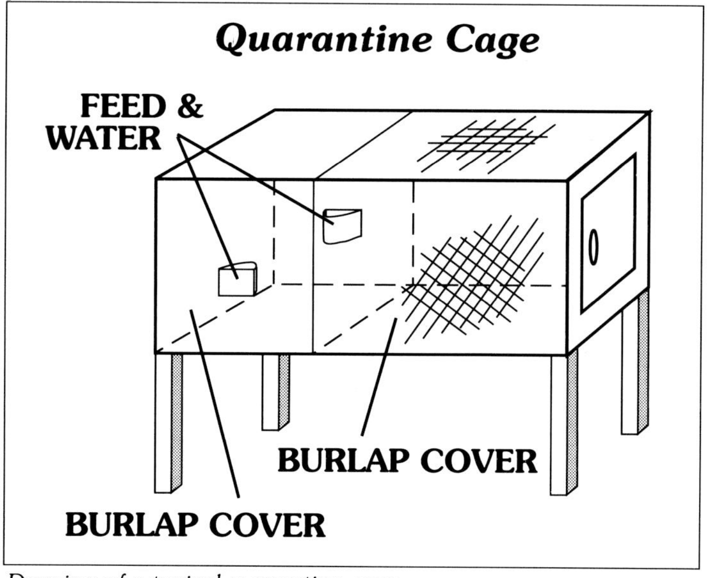

Drawing of a typical quarantine cage.

Such a cage should be elevated with a wire bottom. Elevated feed and water containers should be serviceable from the outside. Burlap protection on the walls is ideal to minimise crashing injuries due to fright. A large enough entry will enable easy netting. Sentinel birds may be placed beneath the cage.

3. Stress and disease

Stress makes birds more susceptible to disease. All species in this group, except perhaps the Red-rumps and the Bourke Parrots, are very susceptible to stress. Stress can come in the form of noise, from predators such as mice and rats at night, from an incorrectly designed aviary, from unhygienic aviary management, from the incorrect diet during the breeding season or anything which affects the physical and psychological well-being of the bird. Stress control is a very important part of disease prevention.

4. Recognising disease

Early recognition of disease is extremely important in their control. These smaller parrots succumb to the effects of disease more quickly than their larger relatives. Neophemas are particularly sensitive to the effects of disease due to their relative inability to conserve body heat. Get to know your birds' behaviour. Any changes which may signal illness will then be noticed earlier. Initial signs of an unwell bird are often subtle.

Watch for a bird: less active than normal, fluffed up with both feet on the perch, rubbing its head on the perch, with a slight tail bob (this could also indicate hens about to lay), with its head behind its wing, sitting low on the perch, which appears sleepy, not eating with the other birds, which does not fly away when you enter the cage, lethargic on the ground.

Or the bird may: sneeze, have wet feathers around the head, have a wet eye, have a shut eye, have a droopy wing, have faeces stuck to the vent or tail feathers or be unable to fly.

Once recognised as being ill, the bird should be moved to a specially designed hospital cage (details page 77). This cage provides an opportunity for the bird to rest in a controlled environment, isolates the bird from the rest of the aviary and allows you to more closely observe the bird and to start any intensive treatment which may be necessary.

5. Emergency procedures in the event of disease

There are a few basic emergency steps to follow which may well save the life of the sick bird:

(a) Prepare the hospital cage. It is necessary to place the ill bird under the least amount of stress as possible. The bird should be caught in the aviary with a net and held no more than is necessary before being placed into the prepared hospital cage. Examine the bird and note any symptoms which may help the veterinarian make his diagnosis. The cage and feed/water containers should be already disinfected and filled with fresh food and water. A piece of clean newspaper or, better still, "Gladwrap" should be placed on the floor in order to collect a fresh dropping sample for analysis. ("Gladwrap" is only used initially to collect a good dropping sample).

(b) Provide warmth. Keep the temperature of the hospital cage at 80-90F or 28-30C. The designed hospital cage will have a thermostatically controlled temperature (see Toby Martin's hospital cage) If you do not have such a cage, place the bird in a cage and provide heat by a heat lamp or by placing a 40-60W light bulb about 10cm from the cage. Check the temperature often. Birds that pant or hold their wings away from the body indicate that the temperature should be lower. The temperature should be increased if the bird remains fluffed.

(c) Provide food and water. Keep food and water within easy reach of the bird. Move the food or rearrange the perches so that food is constantly in front of the bird. Offer favourite and high calorie foods, try sprouted seeds etc. Never remove the bird's normal diet and always provide fresh unaltered water. (Drugs and vitamins in warmed water will deteriorate rapidly and in fact favour bacterial growth). Dehydration is the sick bird's worst enemy. Offer fruits and vegetables with high water content.

(d) Observe that the bird is eating. Crop feeding may be necessary if the bird is not eating and appears weak. The technique is described below, but it should be demonstrated to you by an experienced person. Crop feeding may be the only hope for many sick Neophemas and Grass Parrots which dehydrate much more quickly than the larger parrots. Severely dehydrated birds often develop permanent kidney damage. It is very important in this group of parrots to crop feed too early rather than too late. Crop feeding provides both fluids and nutrition to the ill bird. As well, medications can be given more effectively with this technique.

(e) Collect a dropping and phone your avian veterinarian for advice. He/she may recommend some immediate therapy whilst the droppings are analysed. It is important to collect a fresh dropping and wrap it in an airtight manner before any medication is given. It is difficult to identify many diseases once medication has been given.

(f) Provide circadian light. During normal daylight hours do not cover the cage because birds will not eat in darkness. A night light may even stimulate a sick bird to eat.

6. First aid for the ill bird

When the bird shows signs of illness the situation must be treated as an emergency. Because of its size, the ill bird's temperature drops rapidly once it stops eating and together with dehydration sends the bird into circulatory failure and kidney shut-down. Birds rely on body heat, rest and hydration to overcome illness. The most important first aid steps are mentioned in section 5.

It is a good idea to learn a force-feeding technique utilising a crop needle. Most bird clubs have demonstrations on these techniques. A good energy mix for the ill bird is glucose and water (1/4 teaspoon of Glucodin powder to 30mLs of water) given warm by crop needle or placed in the water dish. This solution must be made fresh and replaced within 12 hours.

7. How to get the best out of your avian veterinarian

The bird patient is unique in medical science. No other animal masks the signs of illness so well. It is thought that birds try to hide symptoms of disease due to their natural fear of being attacked by members of their own flock, since the flock does not want any member attracting predators. It is this innate ability to hide the signs of illness which makes the job of the avian veterinarian so challenging.

A quick and accurate diagnosis is essential if the bird patient is to be saved. The diagnosis of most bird conditions can be made more quickly now that there are reliable diagnostic procedures available (eg. blood tests, bacterial culture and special stains used on body fluids.) However, an accurate diagnosis can be made difficult if incorrect first aid treatment is given before the veterinarian is consulted.

The flock-health takes precedence over the health of the individual in the aviary bird situation and sometimes, sad as it may seem it becomes necessary to sacrifice the life of one ill bird in order to save the rest of the flock.

To get the best out of your avian veterinarian follow the guidelines below.

(A) First assess your disease situation

If one or at most two birds fall ill then it is more than likely that there is no serious threat to your aviary. Immediately isolate these ill birds and treat them with appropriate first aid. Contact your avian veterinarian for further advice. It is important to observe the flock carefully over the following two to five days and if further illnesses occur than it is almost certain you have an infectious problem and ill birds should be submitted as soon as possible to the veterinarian for immediate analysis.

(B) Submitting birds to the avian veterinarian

To expedite the correct diagnosis it is imperative that ill or dead birds taken for analysis have not been treated with any medication. It is best to submit an ill live bird so that the symptoms can be observed and tests taken. Some diseases are difficult to detect in the dead bird.

Place clean newspaper or "Gladwrap" on the floor of the transport cage so that droppings can be collected and tested immediately on arrival at the Veterinary Hospital. Autopsy cannot help in making a correct diagnosis if the bird has been dead for more than six hours. It is important, therefore, to preserve them as soon after death as possible. The dead bird must be cooled if the organs are to be preserved for autopsy. This is best achieved by wetting the bird in detergent (this prevents the feathers from insulating the body), wrapping the body in paper and then plastic. The body is best stored in the vegetable crisper part of the fridge if you are able to get to the veterinarian within four days. If it must be longer then it is best kept in the freezer section. Transport the body to the veterinarian in a cooler.

A detailed autopsy is sometimes the only way of diagnosing a disease in the aviary. Only a few conditions can be accurately diagnosed by looking at the organs alone. It is for this reason that a detailed autopsy is usually required. For the exact treatment recommendations the organs must be sent to an avian pathologist for microscopic analysis, with the results back in about 48 hours.

8. Control and prevention of disease

It is necessary for you to know what diseases are present in the aviary before a reliable disease prevention programme can be initiated. For this reason any ill or dead birds should be properly analysed. As well, a knowledge of those diseases most common to this group will help to prevent diseases entering the aviary. At the time of purchase you can check for these diseases, and treatment can be given in quarantine for those diseases difficult to detect by examination alone. The health of expensive new birds can be thoroughly checked by your avian veterinarian.

Neophemas and Grass Parrots, being ground feeders, are susceptible to diseases spread via droppings - notably worms (roundworms, hairworms and gizzard worms), chlamydiosis and bacterial enteritis. The Psephotus Grass Parrots eat insects and even gastropods (snails/slugs) which can give tapeworm or proventricular worms. Breeding diseases such as French Moult (Psittacine Beak and Feather Disease PFBDS) papovavirus and megabacteria do not escape this group. These diseases are difficult to detect outside the breeding season. External parasites (such as feather mites, feather lice, red mite, cnemidocoptes) are a problem with the smaller Neophemas (particularly the Turquoisines) rather than the larger Grass Parrots.

The following list outlines the diseases which must be checked for on purchase or treated for during quarantine:

At the time of purchase, check for:

1. EXTERNAL PARASITES Examine the primary flights and tail feathers close to feather follicles using a pen-light. Examine the beak, cere and vent for signs of scaley face mite (appears like whitish mould).
2. CHLAMYDIOSIS The signs of Chlamydiosis, especially in Bourkes, Turquoisines and Red-rumps are watery eyes, feather loss around eyes and nostrils, abnormal shaped nostrils and "going light".
3. WORMS A bird eating excessively and "going light".
4. MEGABACTERIA A bird eating excessively and "going light".
5. PAPOVAVIRUS AND PFBDS Abnormal feathers. In Bourkes, Turquoisines, Golden-shoulders and Hooded Parrots, look for loss of tail feathers, deformed tail feathers and loss of feathers around head. (These signs also appear with an exceptionally heavy moult).

In quarantine, check for:

1. Mites and internal parasites using Ivermectin. (Available by prescription only).
2. Chlamydiosis - using Aureomycin (antibiotic) according to the instructions of your avian veterinarians. (Available by prescription only).
3. Other external parasites - treated with Carbaryl powder (from Pet Shop).

Prevention of disease within the aviary

The Neophema and Grass Parrot's year can be divided into the non- breeding, moulting and re-caging seasons. The disease prevention programme is designed around these seasons. The time of onset of the breeding and moulting seasons may vary from year to year. However, usually for this group (excluding the Golden-shoulders and Hooded Parrots) the breeding season begins in late July/early August, and the moulting occurs around January/February. The Golden-shoulder and Hooded Parrots may breed in August/September or in February depending on climatic conditions, and moult around Christmas/late December and again after Winter (July). Most bird keepers purchase new birds in late February (the re-caging season). It is best not to use antibiotics and other medications during the moulting and breeding seasons, and special attention should be given to the recaged bird.

Some diseases in this group occur more often in one season than the other (eg. worms occur after wet/humid weather). Chlamydiosis occurs more often during the re-caging season. PFBDS becomes more visible after the moult. Your vet formulates prevention programmes taking this into account.

9. Worms and a worming programme

Worming is best done using the smallest size crop needle and a tuberculin (1mL) syringe. The small calibrations of the tuberculin syringe gives a very accurate measurement of the small volumes of worm medicine needed for these small species. It is better to use smaller volumes so that reflux from the crop into the lungs is prevented. From 0.1mL to 0.2mL dosages are best for the Neophemas and Grass Parrots. (See table for individual dose rates).

The worm medicine given in the water is less reliable because Levamisole is bitter tasting. Panacur 2.5 settles out quickly and therefore needs regular stirring. Droncit and Ivermectin are not soluble in water. Neophemas and Grass Parrots do not drink much water.

Dosing directly into the mouth is not recommended because it is difficult to give an exact dose and there is a great chance of swallowing the worm medicine into the lungs if a volume of greater than 0.1mL is given.

Some studies have indicated that Ivermectin given by a "spot on" technique (ie applied onto the skin on the back of the neck) is reliable, and this may be the preferred method of worming for those not expert in the use of a crop tube. It is necessary to worm Neophemas and Grass Parrots at least three times a year. There will be deaths from obstructed and ruptured bowels if this group of birds is not routinely dewormed. The most appropriate times to deworm your birds are before the breeding season, after the breeding season and after the warm humid weather seen around Christmas. If there has been a previous problem with worms it may be advisable to worm the fledglings.

In summary

Use a crop needle to dose against worms. Worming Programme: A. Worm one month before the breeding season (1st July) with Panacur or Ivermectin and repeat in three weeks with Levamisole (Nilverm). B. Worm the parents and babies at time of fledging using Panacur, Ivermectin or Levamisole. C. Worm the parents and the young birds at the end of breeding season. D. In aviaries where a worm problem has been diagnosed it is necessary to repeat the worming process in early December and even repeatedly each three months throughout the year.

Below is a list of the recommended dose rates for each species of Neophema and Grass Parrot. See disclaimer. It must be noted that some species can vary in their average body weight and the dose must be adjusted up or down accordingly. Ideally, the bird should be weighed before dosing to ensure the correct dose, but the following list can be used as a guideline when using safe medicines such as the wormers listed. There appears to be greater variation in the weight of the Red-rumps and Bluebonnets (red vented) than for the other species weighed in my survey. The medicine should be measured using the tuberculin syringe so that an accurate dose of anthelmintic can be given. The "spot on" Ivermectin is measured and given with a tuberculin syringe without the needle attached.

Dose rates for deworming each species

Species	Sex	Av Weight	Name & Volume of Anthelmintic		
			Ivermectin	Panacur 2.5	Levamisole
			(1:100 dilut.) dose 0.2ml/100g Spot on Ivermectin (1:25)	dose 0.2mL/ 100g in millilitres 0.08mL	dose 0.1mL/ 100g
Blue-wings	M	45-50g	1 drop	0.09-0.10mL	0.05mL
	F	45-50g			
Elegants	M	45-50g	1 drop	0.09-0.10mL	0.05mL
	F	45-50g			
Turquoisines	M	45-50g	1 drop	0.09-0.10mL	0.05mL
	F	45-50g			
Scarlets	M	45-50g	1 drop	0.09-0.10mL	0.05mL
	F	45-50g			
Bourkes	M	45-50g	1 drop	0.09-0.10mL	0.05mL
	F	45-50g			
Rocks	M	50-55g	1 drop	0.10-0.11mL	0.06mL
	F	55-60g		0.11-0.12mL	
Hooded	M	60-65g	2 drops	0.12-0.13mL	0.06mL
	F	55-60g		0.11-0.12mL	
Golden-shoulder	M	55-60g	2 drops	0.11-0.12mL	0.06mL
	F	50-55g		0.10-0.11mL	
Red-rumps	M	65-100g	3 drops	too variable	0.07mL
	F	55-85g			
Mulgas	M	60-70g	3 drops	0.12-0.13mL	0.07mL
	F	55-70g		0.11.0.14mL	
Blue-bonnets	M	100-130g	4 drops	too variable	0.1mL
	F	100-140g			

Note Carefully: There have been no scientific studies documenting the safe use of these anthelmintics in Neophemas and Psephotus Grass Parrots. The use of these drugs remains the full responsibility of the owner of the birds.

The recommended dose rates above have been found to be effective and safe in non-scientific field trials. It is important not to give any other drugs whilst using the above anthelmintics. If there are adverse effects contact your avian veterinarian for advice.

Ivermectin is available only by prescription. Note the spot on treatment is 1:25 dilution whilst the oral dosage is 1:100. Panacur is 2.5 strength. Levamisole is the 1.4% w/v Nilverm Pig and Poultry wormer.

10. The best method of dosing and admin.

It is important to know the weight of the birds in order to prevent incorrect dosing. See section 16 but note that some species can vary in their average body weight so weighing is recommended before calculating the dosage rate. The medicine should be measured using a tuberculin syringe to ensure accuracy.

The most useful method of dosing is by crop needle. The advantages over water medication are that the exact dose can be given quickly and with a minimum of stress to the bird. The disadvantages are that the birds must be caught and medicated individually which is time consuming and at times stressful to the birds (especially during the breeding season). However, most Neophemas and Psephotus Grass Parrots drink very little which makes crop tube medication necessary.

Very ill birds may require injections. The injection is usually given in the pectoral muscle and is given under the direction and supervision of a veterinarian. An injection can cause excessive bleeding and endanger the recovery of an ill bird when given incorrectly.

11. How and when to use a crop needle

It is best for someone to demonstrate this technique. The basic procedure is to place the crop needle into one side of the mouth and to pass high to the roof and toward the left of midline of the back of the mouth. Follow the food pipe (oesophagus) down and gently move into the crop. Once in the crop the end of the needle can be felt through the crop wall. If it is in the trachea you will feel no needle tip and the trachea may be bent. Once into the crop the medication or force-feeding mixture is slowly delivered and the crop needle removed. Reflux occurs if the medication is delivered too quickly. The mixture will ooze from the mouth. If this happens, quickly remove the needle and point the bird's head toward the ground so that the mixture falls out and does not enter the lungs.

The size of the crop dictates the volume of mixture or medication that can be given. The adult bird has a much smaller capacity than the nestling. The crop capacity of the adult Neophema is about 1mL and the adult Psephotus Grass Parrot is about 3mLs, whereas a baby Neophema can take 1/4-5mLs of food (age dependent) and the baby Psephotus Grass Parrot about 1/4-6mLs (increasing volume with age up until weaning). The maximum volume to be instilled in the crop of the adult Neophema is 1mL, the nestling Neophema is 5mL, the adult Psephotus is 2mL and the nestling is 6mL. If too much volume is given the medication may enter the lungs and cause pneumonia and death.

The dangers of this method disappear with practice and proper tuition. All clubs should demonstrate this technique to their members. A good practice is to use sterile water until your technique is polished. Sterile water will not cause pneumonia if only a small amount enters the lung.

12. The proper use of drugs

Nowadays we are lucky to have good drugs available and most bird diseases can be treated successfully. It is necessary to know the approximate weight of the bird and dose accordingly using a tuberculin syringe to measure the medication.

No scientific studies on pharmokinetics in Neophemas and Psephotus Grass Parrots have been done and dosage recommendations are based on experience in the field. It is important to follow you avian veterinarian's instructions as to the dose rate as "bigger is seldom better" with drugs for small birds.

13. The more common diseases found

As a group Neophemas and Psephotus Grass Parrots get the same diseases as other parrots. However, there are certain diseases to which the group is more susceptible. Most of this group are ground feeders in the wild and they are therefore more susceptible to dropping-spread diseases in captivity such as worm infestation, Megabacterium and Chlamydiosis. The diseases that will be discussed are by no means the only diseases of this group, but represent the most common diseases encountered. The aim of the aviculturist should be to prevent these more common diseases. There are two categories of disease which must be controlled if the aviary is to be kept disease free. 1. Diseases from outside the aviary. 2. Diseases from within the aviary.

These diseases from outside the aviary represent the most common cause of illness in the breeding aviary. In this group of birds the diseases of greatest concern are Chlamydiosis, Megabacteriosis, Psittacine Beak and Feather (PFBDS), Papovavirus, the redmite and the skin mite "cnemidocoptes pilae".

Chlamydiosis

What is Clamydiosis?

Chlamydiosis is also known as Psittacosis, Parrot Fever and Ornithosis. It is caused by a small, toxic bacteria which is capable of causing death and illness in birds and man.

Which species are most commonly affected? This is the most common disease seen in newly purchased Bourkes, Blue-wings, Elegants, Turquoisines and Scarlets.

What are the signs of Chlamydiosis? The clinical signs of affected birds can vary considerably, however the most consistent sign seen is conjunctivitis which may first be noticed by a wet eye, rubbing of the face on the perch then a thickening of the eyelid of one or both eyes with loss of feathers around the eyes. Other signs include prominent keel bone, fluffed feathers and a soiled vent. In the most severe form the bird may be found dead without any previous signs of illness and in good condition.

How is Chlamydiosis prevented from entering your aviary? There are two rules for the prevention of Chlamydiosis. 1. Only purchase birds with no signs of this disease. 2. Quarantine all newly purchased birds for six weeks.

When purchasing a new bird check carefully for any of the above signs. If any are found, the bird should not be purchased. Unfortunately, the disease may still enter your aviary in birds free from the above symptoms because sometimes, the bird will be a "carrier" of the disease. A "carrier" is a bird who appears healthy but is capable of infecting other birds. A "carrier" bird may become ill when stressed. Changing aviaries is stressful to birds and therefore a "carrier" will often become ill when introduced to a new aviary. If a bird becomes ill during the quarantine period a long course of chlortetracycline will rid the bird of the disease.

Megabacteriosis

What is Megabacteriosis?

This bird disease has become apparent over the past few years in Australia. The bacteria is predominantly found in the proventiculus (stomach) and gizzard. The bacteria changes the acidity of this part of the bowel and appears to prevent adequate absorption of nutrition.

Which species are most susceptible to Megabacteria? Of this group, the Elegants, Rocks and Blue-wings and especially their colour mutations are most susceptible to the effects of this disease.

What are the signs of Megabacteriosis? The most common signs of this illness are going light, a watery pale coloured dropping and a fluffed up, poor, dirty plumage. However, like

Chlamydiosis "carriers" occur.

Can Megabacteriosis be treated? Antibiotic sensitivity has revealed that the germ is completely resistant to all therapeutics commonly used. No specific therapy is known, but some positive effects are seen after feeding only a soft food mix and grit, together with drugs to counter any secondary infection.

How is Megabacteriosis prevented from entering the aviary? The same rules as for Chlamydiosis apply for Megabacterium. Select only healthy birds, showing none of the above signs and quarantine for a period of six weeks. At the beginning and end of the quarantine period, a dropping sample should be examined for Megabacterium. Any birds showing any symptoms should not enter your aviary. There is little chance of a vaccine being developed against Megabacterium so a cautious approach to this disease will help keep your aviary disease free.

Psittacine feather and beak disease (PFBDS)

What is Psittacine Feather and Beak Disease?

Feather and Beak Disease is most well known from the syndrome seen in the Sulphur-crested Cockatoo. It is a virus infection affecting the feathers and beak in the Cockatoo. However, in the Neophema/Psephotus Grass Parrot group the same virus causes a feather and nestling disease with no beak involvement.

Which species are affected?

The disease is seen in all of the group, both in the wild and in captivity. It has even been found in the Orange-bellied Parrot, the endangered species of Tasmania. It is most prevalent in the Scarlets, the Turquoisines, the Bourkes, the Golden-shoulders and Hoodeds (and colour mutations of these species).

What are the signs of PFBDS?

In this group, the most outstanding sign of the disease is the loss of tail and head feathers at the time of feather growth or the first moult. It is usually a young bird disease but does occur in older birds usually at breeding time or during the moult.

How is PFBDS prevented from entering the aviary?

Birds in poor feather or without tail feathers should not be purchased. Quarantine does not help prevent this disease as it will not show up until the next moult or breeding season. "Carrier" birds do occur, so it is important to buy your birds from an aviary known to be free of this disease.

Is there a treatment for PFBDS?

Because the disease is caused by a virus there is no known treatment at the present time. A vaccine to protect from this disease is not available.

Skin Mites (Cnemidocoptes)

What is the skin mite?

The skin mite Cnemidocoptes is best known as the cause of "scaley face" in Budgerigars and "tassle foot" in Canaries. It is a microscopic mite which lives in the skin causing irritation and feather loss.

Which species does the mite affect? It is usually seen in the Turks, Scarlets and Bourkes. It tends to be present in particular families of bird species and may be related to an inherent weakness of these families to mite infestation.

What are the signs of skin mite?

In this group, the skin around the vent and tail butt is most commonly affected. The skin will be scaly, thickened and there may be sores from self picking. The bird will over-groom in the areas of infestation.

What is the treatment for this mite?

Most birds will respond to veterinary treatment with an Ivermectin 1% solution over a period of several weeks, however, a bird with skin mite has an inherent weakness to this mite and should not become part of your breeding programme. This mite can be confused with the feather mite which is a reflection of bad management alone and is of no harm to your breeding establishment once it has been treated in isolation.

How is the skin mite prevented from entering the aviary?

The skin mite can be prevented only by purchasing birds which are not over-grooming nor have signs of skin scale, especially around the vent.

The basic Rules used to prevent disease entering an aviary

1. Purchase healthy and alert birds only.
2. Do not purchase during the moult or in poor feather or with tail feathers missing, with watery eyes or going light.
3. Quarantine all newly puirchased birds for 6 weeks in an isolation cage.
4. Observe for illness whilst in quarantine.
5. Ill birds in quarantine must be diagnosed and treated appropriately before entering the aviary.
6. Only healthy quarantined birds are to enter your aviary.

Good management prevents disease within the aviary. Good management takes into account the system of housing and quarantine, aviary hygiene, breeding preferences, proper nutrition, accurate record keeping and an informed policy against disease. By providing the best physical and psychological environment for the captive bird, breeding results are maximised and disease minimised. The above aspects of good management have been well covered by Toby Martin. I will discuss the diseases which most commonly occur within the aviary.

The diseases of note are worm infestations, Papovavirus and egg binding. See section 8 for a worming programme. Quarantine helps in preventing the entry of most diseases but it can never be 100% effective. Egg binding occurs because of nutritional and sometimes hereditary factors.

Roundworms

Roundworms are not the only worms causing a problem, however they are the major worm seen in these species. This section deals with the control of roundworm.

How are Roundworms controlled?

Whenever a worm infestation occurs, the management must be reassessed and tightened. Control of recurrent problems can be reduced by modern aviary construction with birds housed off the ground, or on readily cleaned concrete floors. Concrete floors are easy to clean and disinfect but may hold moisture at times. Eggs must undergo development in the environment before becoming infective to the next host. Warm and wet conditions enhance this development. Therefore regular scraping and a dry aviary are of utmost importance to prevent the spread within the aviary. Clean aviaries mean healthier birds. Also, overcrowding may effect the health of the birds. In an aviary with a dirt floor, if a roundworm problem is diagnosed, the only way of removing the eggs is by removing the top 10cm of soil. Never dig it up since this will protect the eggs. The best way to kill eggs is with a blow torch. Disinfectants will not kill the eggs in soil. Caustic soda will kill the eggs along with all the vegetation.

Rules for the control of Roundworms

1. Control entry of worms. Treat all birds before entering aviary.
2. Treat birds in aviary with an appropriate anthelmintic at the appropriate times (See section 8.)
3. Prevent the worm egg from larvating by maintaining dry conditions, removing droppings within 48 hours or by using elevated wire cages. With dirt floor aviaries, regular removal of the top 10cm of soil or treatment with soda lime especially after wet weather may be necessary.
4. Prevent the stress of captivity so as to lessen the effects of a worm infestation. Ensure all birds are on a good balanced diet, especially the young birds, do not overcrowd and prevent inter-species aggression.

Papovavirus

What is Papovavirus?

Papovavirus is a breeding disease seen mostly in nestlings and fledglings.

What are the signs of Papovavirus?

Papovavirus may appear in several forms. Usually there will be sudden deaths in nestlings. In older birds there may be feather disease similar to PFBDS.

What is the treatment for Papovavirus?

There is no effective drug against this disease. If there is a suspicion of Papovavirus, an avian veterinarian is necessary to confirm the disease and to advise you on a control and prevention programme.

Egg Binding

What is egg binding?

Egg binding is the inability of the hen to lay the formed egg. There are many causes of egg binding but the most common is obesity or inadequate calcium in the diet. Neophemas and Psephotus Grass Parrots are active birds and obesity is rare in an adequate sized aviary, however, certain families of birds may be more susceptible to egg binding so the most likely cause is a dietary deficiency in calcium.

What are the signs of egg binding?

The signs are sudden in onset in an otherwise previously healthy breeding bird. The first sign of a bird in trouble may be that she is sitting fluffed up on the bottom of the cage. She may be unable to fly and may have abandoned her eggs. There is a lot of pain with the retained egg placing pressure on the sciatic nerves, the kidneys and ureters. This can be a life threatening situation and treatment must be swift.

What is the treatment for egg binding?

Heat (80-85F) and humidity is the immediate treatment, together with calcium drops by mouth (0.1mL of Calcium Sandoz

Syrup each hour) until she lays the egg. If she is in distress and fails to pass an egg within three hours you should seek veterinary assistance.

How is egg binding prevented?

Prevention is achieved by maintaining a fit bird on a good diet, in a warm environment. Calcium Sandoz Syrup added to the drinking water before egg laying may help prevent the birds who have previously been egg bound.

Informed disease control

Whenever a bird becomes ill or dies it is important that the disease is accurately identified, so that the disease status of your aviary is known. This allows your avian veterinarian to formulate a disease control programme to protect the rest of the aviary at this time and in the future. Good record keeping allows the disease process to be understood quickly so that the spread can be halted rapidly.

To prevent a disease spreading through the aviary, regular surveillance for ill birds should be part of a good management programme. The ill bird should be isolated into a warmed hospital cage and the disease identified by an avian veterinarian before any medication is prescribed. Indiscriminate use of drugs is dangerous as it makes diagnosis of the disease difficult and may lead to resistance of the disease organisms to certain antibiotics. Any dead birds should be autopsied to determine the cause of death so that appropriate treatment can be started.

14. Notes on some diseases affecting babies

Rickets
(Lameness, malformed limbs and inability to fly)

The most common disease of young Grass Parrots especially the Red-rump. Rickets is caused by too low levels of calcium in the diet of the parents or hand rearing mix. During the breeding season it is necessary to give more calcium in the form of cuttlefish or a calcium additive in the water. An avian veterinarian should be consulted if rickets occurs on a diet high in calcium because there will be another disease affecting the absorption of calcium.

Vomiting

Any baby bird which becomes ill and vomits should be immediately isolated from the other handreared babies. The most common cause of vomiting in handreared birds is Thrush (Candidiasis). This is a disease associated with improper hygiene or may occur second to another disease. Thrush is treated with Nystatin (available by prescription only). There is probably another disease present if the condition does not respond to Nystatin after two days. Papovavirus, worms, Megabacteria or food which is fed too cold are other causes of vomiting. If vomiting continues, an avian veterinarian should be consulted.

Stunting

Stunting of hand reared babies occurs when the feed mix is too watery (malnutrition), the protein is too low or too high. If the diet is adequate, there is another disease present and an avian veterinarian should be consulted. A virus disease such as Papovavirus or PFBDS may be involved if stunting is associated with abnormal feathers.

15. Some reference points for hand rearing

The Psephotus Grass Parrots and Neophemas grow extremely quickly. The first three days is the most critical stage in the life of any parrot and especially so for the very small Neophema Parrots. During these first three days, the baby bird requires about 25% of its own body weight of food a day to develop normally. This must be given in a 93% water suspension (7% solids which do not settle out) at 40 degrees centigrade each two hours for these first three days of life. After three days feed 30% solids until weaning. The best way to calculate 30% solids is that it is about the consistency of melted ice-cream or thin yoghurt.

A baby Bourke Parrot is fed from a "bent" teaspoon

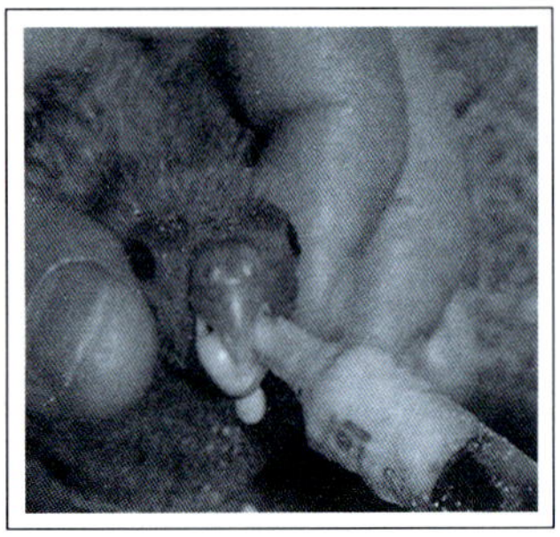

A baby Turquoisine is eating from a dyalisis syringe

The ingredients of the diet can be varied and Toby Martin's recipe (found in the hand-rearing section at the end of the book) is an example of a good diet. Theoretically the diet should contain 20% protein in a cereal base, at least 4% fat, and a good balance of vitamins and minerals. There should be adequate but not excessive vitamin additives and a calcium supplement of 2% dry matter of the diet.

16. Surgical Sexing

This is a procedure performed under a gas anaesthetic which allows birds to be sexed at a young age, giving a breeder an opportunity to select definite pairs before the breeding season. It requires an avian veterinarian to perform the procedure. A very small fiber optic light (endoscope) is inserted into the abdominal cavity through a small incision and the sex organs are visualised. Blue-wings, Elegants, Rock Parrots and Blue-bonnets are the parrots of this group most often sexed. During the procedure other organs are examined for abnormalities. The birds breeding capability can also be determined. Neophemas are usually sexed at about 4 months of age.

****

Double hospital cage and brooding box as used by the author

Brooding Box

The total unit measures 1.04m long x 43cm high x 33cm deep and is made of laminexed chipboard for easy cleaning. Each cage is 43cm long leaving a width of 18cm in the middle to house the heat source. The roof of the unit is 2.5cm less in width than the base. This allows the perpendicular fitting of plastic tracks to allow operation of the front sliding covers.

One cage is used as a baby chick unit, the other houses older chicks to be weaned onto seed. The baby unit is thermostatically controlled. The warm air enters through four vent holes in the side. The temperature is maintained at 30-32C. The babies are kept in small boxes (as single birds or groups - from one nest - as required). A middle wire shelf allows room for two tiers of boxes.

The older baby unit does not contain a thermostat. Only one vent allows warm air to enter. The temperature in this unit is about 28C. There is a tin tray at the bottom to facilitate cleaning out. Seed, water dishes and perches are also provided. Both contain a minimum-maximum thermometer. Both units have wire cage fronts with an outside sliding perspex front.

The motor housed between the units is an intake fan unit from an incubator. Heat is provided by two 60 watt globes. The 4cm hole at the bottom is covered in a fine gauze to filter the air. A 5cm x 10cm perspex hole allows checks to be made on the globes.

The hospital cage unit is identical. Unlike many hospital cages, this version does not force the bird or food and water dishes to be directly above the light bulb. The temperature is thermostatically controlled and is kept at about 35C.

Handrearing Baby Birds

Sometimes babies are not fed by the parents as well as they could be. Perhaps it is a bigger clutch than usual or at worst the hen has died. For whatever reason, babies, especially the rarer mutations, should not be simply written off. Handrearing is a necessary function that should be mastered by all aviculturists. I will give you the mixture's recipe at the end of this section.

The amount of mixture required at each feed (obviously depending on the number of baby birds), should be placed in a small dish - a tea cup is ideal. A little cold water is added first and the mixture is stirred into a smooth paste - this avoids lumps. Boiling water is then added from the kettle and stirred in to the consistency required - a yoghurt consistency is fine. It should run fairly easily off the spoon. It must be fed fairly hot to the birds - bearable to the inside wrist or top lip. Refusal to feed often indicates a temperature problem. The small dish or cup is placed in a bowl of hot water to keep it warm enough for the last bird. Boiling water is added to the big bowl to heat up the food if required. The amount of food taken may vary from bird to bird. The crop should be filled but not to the extent where it is stretched to the limit. Common sense is all that is required. If the baby is over ten days old the crop should be practically empty prior to each feed. Four feeds a day should be quite sufficient. The first feed between 6am and 8am, the second between 1pm and 2pm, the third between 5pm and 6pm and the final feed before you go to bed between 10pm and midnight. The birds will make it through the night without a feed. Less than ten days old, feed when the crops are empty. This could be anywhere between two and five hours apart. Again, even at this age, a feed as late at night as possible and one early in the morning will be alright. 3am feeds are rarely required.

My wife, Jacki, finds the easiest and cleanest method of feeding is with a 2mL dialysis syringe. The food is sucked up - don't put it in the top of the syringe, as what goes in the top won't necessarily get through the narrow exit!. Place the tip of the syringe into the side of the mouth and gently squeeze the food in. The baby will swallow it normally. Do not feed directly into the crop. Another successful method is to use a teaspoon bent up at the sides to form a funnel. Both methods are shown quite clearly in the photographs. Wipe the beaks and, if necessary, face and throat feathers clean of food after each feed. One of the main reasons our babies are kept in a warm brooder is that if a bird with warm to hot food in its crop is kept in a comparably cool place, Sour Crop problems can occur. Since using the brooder, Sour Crop has never been an issue.

Should you have an outbreak of Sour Crop, the following steps should be taken:

If Sour Crop is forming put 3 drops of Mycostatin in with the food and massage the crop.

If it has seized up completely use Mycostatin in water, no food until the crop is clear. Once the crop is clear, feed natural yoghurt and stewed apple with a pinch of Glucodin for two to four feeds and then go back to a runny diet with 1 drop of Mycostatin added per feed.

The following recipe is one that I have used with great success. There are many such recipes available, find one with which you are successful and stick to it.

MIX A - For birds up to pin feathers 4 oz High Protein Baby Cereal 4 oz Egg and Biscuit Mix 2 oz full Cream Milk Powder 1 heaped desertspoon of: Maize Meal, Ground Rice, Arrowroot Biscuit and Millet meal (all four mixed in equal quantities).

For New Born Babies - 1 meal daily of plain natural yoghurt mixed with stewed apple (Blue Heinz) for the first three days.

MIX B - for older babies 8 oz High Protein Baby Cereal 2 oz Sunflower Meal 4 oz Crushed Arrowroot Biscuit 4 oz Raw Wheatgerm 1 teaspoon Glucodin 1 heaped dessertspoon of: Maize Meal and Millet Meal (both mixed in equal quantities) + Pentavite Vitamins - 1 drop per bird once a day

Genetic Tables

It would obviously take up too much space to write out the possible matings and expectations of the various mutations mentioned in this book.

Apart from the Jade and Olive Turquoisine's dark factor inheritance, already shown alongside these birds, all mutations mentioned are either Recessive or Sex-linked.

These modes of inheritance are shown in a basic table form below:

Recessive Inheritance

NN = A normal bird cc = A recessive coloured bird

Nc = A normal, split to a recessive colour

Examples of recessive colours are:

Full Red-fronted Turquoisine, Yellow Turquoisine, all three Blue Scarlet-chesteds, Blue Red-rumps.

NN x NN = 100% NN NN x cc = 100% Nc Nc x cc = 50% Nc 50% cc Nc x Nc=50% Nc 25% NN 25% cc

cc x cc=100% cc

Substitute 'c' for the recessive colour.

Sex-Linked Inheritance

Examples of sex-linked colours are:

Rosey and Pink Bourkes, Cinnamon, Elegants, Lutino Red-rumps.

S.L. = A sex-linked coloured bird Non S.L. = A bird that is not a sex-linked colour Non S.L./S.L. = A non sex-linked coloured bird that is split to a sex-linked colour

Cock Hen Progeny

S.L. x S.L.=ALL S.L. (both sexes)

S.L. x Non S.L.=Cocks: Non S.L./S.L. Hens: S.L.

Non S.L. x S.L.=Cocks: Non S.L./S.L. Hens: Non S.L.

Non S.L./S.L. x S.L.=Cocks: Non S.L./S.L. & S.L. Hens: Non S.L. & S.L.

Non S.L./S.L. x Non S.L.=Cocks: Non S.L./S.L. & NonS.L Hens: Non S.L and S.L.

Substitute S.L. for the sex-linked colour.

Note when two different sex-linked colours are bred together, the hen acts as a non sex-linked bird.

For example in the Red-rumped Parrot a Lutino cock mated with a Cinnamon hen would produce normal cocks split to Lutino and Cinnamon and Lutino hens.

A word of warning

There is no doubt that there will be increases in the number of mutations in the birds discussed in this book. Established colours will become cheaper and more widely spread. Let us make sure that we do not lose our lines of pure normal birds. Not only must the original birds be maintained for their own sake but also for the fact that many mutations, especially newer ones, must be bred back through good normals. This ensures that mutations remain healthy, vigorous and fertile and also conform to the standard of the normal bird.

Let us hope we have learned something in this regard from the African Lovebird mutations and not repeat any of the mistakes of mass producing inferior birds.

We can start now by rectifying our first mistake which reduced the Yellow Turquoisine to a weedy little bird. This was a prime example of what harm the "dollar" can do to birds in certain peoples' hands.

Ex-Lovebird breeders are now turning to Grass Parrots and their mutations. If some of these breeders go for quantity again rather than quality then buyers should be adamant and not purchase inferior birds. The breeders who worship the dollar rather than the bird will only change their ways if the market dictates it.

▲The author's double brooder

▼The author's twin hospital cage unit

▲*The hot-air source of the brooder and hospital units*

Publishers Note

This title is published by Australian Birdkeeper Pty Ltd who produce a wide and varied range of avicultural literature including the world acclaimed Australian Birdkeeper magazine - a full colour, 48 page bi-monthly magazine specifically designed for birdlovers and aviculturists. It is the intention of the publishers to produce high quality, informative literature for bird lovers, fanciers and aviculturists alike throughout the world. It is also the publishers' belief that the dissemination of qualified information on the care, keeping and breeding of birds is imperative for the total well-being of captive birds and the increased knowledge of aviculturists.

Nigel Steele-Boyce.

Nigel Steele-Boyce
Editor-In-Chief
Australian Birdkeeper